Analytical Mathematics Level 2

Analytical Mathematics Level 2

ERIC WALKER, MA(Cantab.)

Formerly Head of the Mathematics Department and Deputy Headmaster of Sir Roger Manwood's School, Sandwich; later at the South Kent College of Technology, Folkestone.

HOLT, RINEHART AND WINSTON

LONDON · NEW YORK · SYDNEY · TORONTO

Holt, Rinehart and Winston Ltd: 1 St Anne's Road,
Eastbourne, East Sussex BN21 3UN

British Library Cataloguing in Publication Data

Walker, Eric
Analytical mathematics level 2
1. Shop mathematics
I. Title
510′.246 TJ1165

ISBN 0-03-910343-9

Typeset by Macmillan India Ltd, Bangalore.
Printed in Great Britain by Mackays of Chatham Ltd, Chatham, Kent.

Last digit is print no: 9 8 7 6 5 4 3 2 1

Introduction

eal number

. real number is a number whose square is not negative. That is, where n is :al, then

$$n^2 \nless 0, \quad \text{or} \quad n^2 \geqslant 0$$

unction

his term is used frequently both in the TEC syllabus and throughout this ook, but at previous levels there has been no formal definition of this term. ecause many students do not have a clear understanding of 'function' they se the word loosely, in circumstances where it is not appropriate. It is sential that the term should be properly understood. There are a number slightly different ways of looking at the matter, and each one has its own lvantage. Together they shed greater light on the subject.

Suppose that x represents a variable which takes all real values, i.e. between $-\infty$ and $+\infty$. Suppose also that $f(x)$ represents some expression, such as $5x^2 - 3x + 2$, $\sin(3x + 5)$, $e^{\frac{1}{2}x^2}$, $\log(5x + 8)$, whose values vary with the values of x. Then $f(x)$ is a function only if $f(x)$ takes on a single value for each and every value of x. That is, $f(x)$ is a 'single valued' expression.

Concept 1 may be interpreted graphically using a mapping diagram, as in Fig. 1. Note that $f(x)$ is a function if and only if each point on x maps onto a single point on $f(x)$.

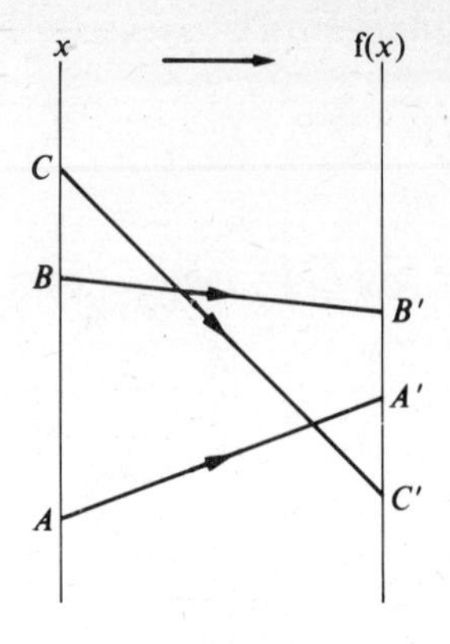

Figure 1

In certain cases a number of different points on x map onto the sam point on $f(x)$. For example, in Fig. 2, other than the value of $x =$ $x = +a$ and $x = -a$ each maps onto $f(x) = a^2$. Nevertheless, x^2 is function because each point on x maps onto only one point on x^2.

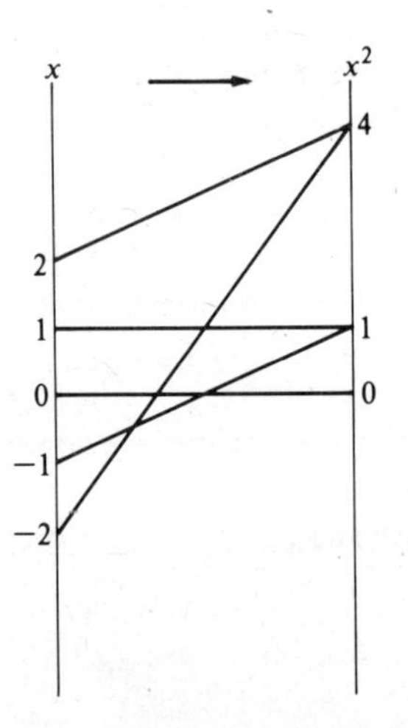

Figure 2

3. A function $f(x)$ is a machine. It accepts a value of x and converts it in another value, $f(x)$. To each x there is just one value of $f(x)$. Compa this with a lift. Think of the values of x as the buttons inside the lift ar the values of $f(x)$ as the floors at which the lift stops. For each button c

the control panel in the lift there is just one position at which the lift stops. Pressing button 2 makes the lift stop at floor 2. A lift which sometimes stopped at floor 2 and sometimes at floor 5 when button 2 was pressed would be of little use. In fact machines in general are particularly important examples of the practical applications of functions. When given an instruction just one particular operation must be carried out by the machine.

Sequence of chapters

The chapter order corresponds to the order in which the sections appear in the TEC unit. It is not necessarily the order in which the chapters should be studied. That is a matter for the teacher to decide. For instance, it might be thought desirable to study Chapter 8 and perhaps even Chapters 9 and 10 too before Chapter 3 is attempted. That would mean that the topics of gradient and differentiation would be introduced before they had to be applied in the earlier chapter.

E. WALKER

To Ian and Anne, Angela and Nick, Glyn and Jayne; my sons and their wives, and my daughter and her husband.

Contents

Unit Reference

Unit BA7:	Chapter 1
Unit BA8:	Chapter 2
Unit BA10:	Chapter 3
Unit BB8:	Chapter 4
Unit BB9:	Chapter 5
Unit BC7:	Chapter 6
Unit BC8:	Chapter 7
Unit CA8:	Chapter 8
Unit CA9:	Chapter 9
Unit CB11:	Chapter 10

1

The Solution of Simple Quadratic Equations Analytically

1.1 The factorization of quadratic expressions

Four termed expressions, by grouping in pairs

$$ax + ay + bx + by = a(x+y) + b(x+y) = (a+b)(x+y) \qquad 1.1$$

Perfect squares

$$\begin{aligned} a^2 + 2ab + b^2 &= a^2 + ab + ab + b^2 = a(a+b) + b(a+b) \\ &= (a+b)(a+b) \qquad \text{by } 1.1 \\ a^2 - 2ab + b^2 &= a^2 - ab - ab + b^2 = a(a-b) - b(a-b) \\ &= (a-b)(a-b) \qquad \text{by } 1.1 \end{aligned}$$

Giving:

$$a^2 + 2ab + b^2 = (a+b)^2 \qquad 1.2$$

$$a^2 - 2ab + b^2 = (a-b)^2 \qquad 1.3$$

1.2 and *1.3* represent the factorization of perfect squares. They will be needed often in later work in this chapter and elsewhere.

The difference of two squares

$$\begin{aligned} a^2 - b^2 &= a^2 - ab + ab - b^2 = a(a-b) + b(a-b) \\ &= (a+b)(a-b) \qquad \text{by } 1.1 \end{aligned}$$

Giving:

$$a^2 - b^2 = (a+b)(a-b) \qquad 1.4$$

1.4 will be needed often in later work.

Examples

1.
$$\begin{aligned} 16x^2 + 40xy + 25y^2 &= (4x)^2 + 2(4x)(5y) + (5y)^2 \\ &= (4x+5y)^2 \qquad \text{by } 1.2 \end{aligned}$$

2.
$$\begin{aligned} 32x^2y^2 - 80xyz + 50z^2 &= 2(16x^2y^2 - 40xyz + 25z^2) \\ &= 2[(4xy)^2 - 2(4xy)(5z) + (5z)^2] \\ &= 2(4xy - 5z)^2 \qquad \text{by } 1.3 \end{aligned}$$

3.
$$\begin{aligned} 49ab^2 - 64ac^2 &= a(49b^2 - 64c^2) = a[(7b)^2 - (8c)^2] \\ &= a(7b+8c)(7b-8c) \qquad \text{by } 1.4 \end{aligned}$$

4.
$$\begin{aligned} a^2 + 2ab + b^2 - c^2 &= (a^2 + 2ab + b^2) - c^2 \\ &= (a+b)^2 - c^2 \qquad \text{by } 1.2 \\ &= (a+b+c)(a+b-c) \qquad \text{by } 1.4 \end{aligned}$$

5.
$$\begin{aligned} x^2 + 2x - 3 &= (x^2 + 2x + 1) - 4 \\ &= (x+1)^2 - 4 \qquad \text{by } 1.2 \\ = (x+1)^2 - 2^2 &= [(x+1)+2][(x+1)-2] \qquad \text{by } 1.4 \\ &= (x+3)(x-1) \end{aligned}$$

Notice that in examples (2) and (3) we detect a common factor of all the terms. In (2) it is 2 and in (3) it is a. In each case the first step is to take out that common factor. This procedure applies whatever the form of the expression. Example (5) provides a preliminary to the factorization of quadratic expressions which are not perfect squares or the difference of two squares.

Formulae *1.2* and *1.3* will be needed frequently for a process which is called completing the square.

Exercise 1.1

Factorize the following:

1. $4x + 4y + ax + ay$
2. $ab + ac + pb + pc$

3. $2ax + 3ay + 2bx + 3by$
4. $x^2 + 4x + 4$
5. $x^2 - 16x + 64$
6. $a^2 - c^2$
7. $2x^2 + 8x + 8$
8. $3x^2 - 48x + 192$
9. $9 - 30a + 25a^2$
10. $pa^2 - pc^2$
11. $9a^2 - 30a + 25$
12. $49x^2 - 28x + 4$
13. $81p^2 - 16q^2$
14. $(2x + 3)^2 - (x - 1)^2$
15. $(x + a)^2 - 4a^2$
16. $(x + a)^2 - b^2$

The factorization of trinomials

A quadratic expression such as $4x^2 - 11x - 14$ is often called a trinomial. A number of different techniques may be used to factorize such expressions.

Method 1

1. $x^2 + (a + b)x + ab = x^2 + ax + bx + ab = x(x + a) + b(x + a)$
   $= (x + a)(x + b)$ by *1.1*
2. $x^2 - (a + b)x + ab = x^2 - ax - bx + ab = x(x - a) - b(x - a)$
   $= (x - a)(x - b)$ by *1.1*

Examples

1. $x^2 + 7x + 12 = x^2 + 4x + 3x + 12 = x(x + 4) + 3(x + 4)$
   $= (x + 4)(x + 3)$
2. $x^2 - 11x + 24 = x^2 - 8x - 3x + 24 = x(x - 8) - 3(x - 8)$
   $= (x - 8)(x - 3)$

Rule. Where the last sign of the quadratic is plus, the signs in the factors are alike. They are actually equal to the sign of the middle term of the quadratic. The numbers in the brackets are such that they add up to the coefficient of the middle term of the quadratic and such that their product is the last term of the quadratic.

3. $x^2+(a-b)x-ab = x^2+ax-bx-ab = x(x+a)-b(x+a)$
$= (x+a)(x-b)$ by *1.1*
4. $x^2-(a-b)x-ab = x^2-ax+bx-ab = x(x-a)+b(x-a)$
$= (x-a)(x+b)$ by *1.1*
5. $x^2+4x-45 = x^2+9x-5x-45 = x(x+9)-5(x+9)$
$= (x+9)(x-5)$
6. $x^2-7x-120 = x^2-15x+8x-120 = x(x-15)+8(x-15)$
$= (x-15)(x+8)$

Rule. Where the last sign of the quadratic is minus the signs in the factors are different. The numbers in the factors have a product which is the last term of the quadratic and a difference which is the coefficient of the middle term of the quadratic.

Method 2

Apply the method adopted in Example (5) in the section on 'the difference of two squares'.

Example

$$x^2+13x+36 = \left[x^2+13x+\left(\frac{13}{2}\right)^2\right]+36-\left(\frac{13}{2}\right)^2$$
$$= \left(x+\frac{13}{2}\right)^2+36-42\tfrac{1}{4} = (x+6\tfrac{1}{2})^2-6\tfrac{1}{4}$$
$$= (x+6\tfrac{1}{2})^2-(2\tfrac{1}{2})^2$$
$$= (x+6\tfrac{1}{2}+2\tfrac{1}{2})(x+6\tfrac{1}{2}-2\tfrac{1}{2}) = (x+9)(x+4)$$

In this particular example Method 2 is more complicated than Method 1, but it does have its advantages in the more difficult problems in the next section. In fact, the method never fails.

Exercise 1.2

Factorize:

1. x^2+3x+2
2. x^2-3x+2
3. x^2+x-2
4. x^2-x-2

5. x^2+5x+4
6. x^2-5x+4
7. x^2+3x-4
8. x^2-3x-4
9. x^2-5x-6
10. x^2+x-6
11. x^2+5x+6
12. x^2-x-6
13. $x^2+3x-10$
14. $x^2-7x-30$
15. $x^2-17x+72$
16. $x^2-6x-72$
17. $2x^2+6x-8$
18. $4x^2+12x-16$
19. $11x^2+33x-110$
20. $ax^2+ax-6a$

Most trinomials do not have a coefficient of $x^2 = 1$. Nor do the terms have a common factor as in questions (17) to (20) in Exercise 1.2. In these instances a certain amount of trial and error is required when Method 1 is adapted to the new circumstances.

Examples

1. $2x^2-7x+3$ is to be factorized.
   Assume $2x^2-7x+3=(A-B)(C-D)$
   Then: (a) $AC=2x^2$; (b) $BD=3$; (c) $-AD-BC=-7x$
   From (a), A and C must be $2x$ and x, or x and $2x$
   From (b), B and D must be 3 and 1, or 1 and 3
   These choices ensure that the product of the factors gives a trinomial with the first term and the last term, $2x^2$ and $+3$ respectively.

   To obtain the proper combinations we need only check that they produce the middle term of the trinomial, i.e. $-7x$. The routine is as follows. First write $(A-B)(C-D)$ in the form:

$$\begin{matrix} A & -B \\ C & -D \end{matrix}$$

   as we do when carrying out a long multiplication in arithmetic.
   First trial:

$$\begin{matrix} 2x & \searrow\!\!\!\!\nearrow & -3 \\ x & & -1 \end{matrix}$$

Here $AD = 2x$ and $BC = 3x$. From (c), $-2x - 3x$ should give $-7x$. It does not. Therefore the combination is incorrect.
Second trial:

$$\begin{array}{ccc} 2x & & -1 \\ & \times & \\ x & & -3 \end{array}$$

Here $AD = 6x$ and $BC = x$. By (c), $-6x - x$ should give $-7x$. It does. This time the combination is correct. Then:

$$2x^2 - 7x + 3 = (2x - 1)(x - 3)$$

2. Factorize $14x^2 + 13x - 12$.
Assume $14x^2 + 13x - 12 = (P + Q)(R - S)$
Then $PR = 14x^2$, $QS = 12$, $QR - PS = 13x$
Possible values of P and R are $14x$, x; x, $14x$; $7x$, $2x$; $2x$, $7x$
Possible values of Q and S are 12, 1; 1, 12; 6, 2; 2, 6; 4, 3; 3, 4
From what has been said it would appear we ought to proceed as follows:

$$\begin{array}{ccc} 14 & & +12 \\ & \times & \\ 1 & & -1 \end{array}$$

Note that we omit the x on the left. Reject: it gives $-2x$, not $+13x$.

$$\begin{array}{ccc} 14 & & -12 \\ & \times & \\ 1 & & +1 \end{array} \quad \text{Reject} \qquad \begin{array}{ccc} 14 & & -1 \\ & \times & \\ 1 & & +12 \end{array} \quad \text{Reject}$$

$$\begin{array}{ccc} 14 & & +1 \\ & \times & \\ 1 & & -12 \end{array} \quad \text{Reject} \qquad \begin{array}{ccc} 7 & & +1 \\ & \times & \\ 2 & & -6 \end{array} \quad \text{Reject}$$

And so on until we reach:

$$\begin{array}{ccc} 7 & & -4 \\ & \times & \\ 2 & & +3 \end{array}$$

Accept, because $21x - 8x = +13x$

This is a lengthy routine. However, there are short cuts. In this case 13 can arise only from an AD and a BC one of which is even and the other odd. So we would never even consider such combinations as:

$$\begin{array}{ccc} 14 & & -12 \\ & \times & \\ 1 & & +1 \end{array} \qquad \begin{array}{ccc} 14 & & +12 \\ & \times & \\ 1 & & -1 \end{array} \qquad \begin{array}{ccc} 7 & & +1 \\ & \times & \\ 2 & & -6 \end{array}$$

$$\begin{array}{ccc} 7 & & +3 \\ & \times & \\ 2 & & -4 \end{array} \qquad \begin{array}{ccc} 7 & & -1 \\ & \times & \\ 2 & & +6 \end{array} \qquad \begin{array}{ccc} 7 & & -3 \\ & \times & \\ 2 & & +4 \end{array}$$

Neither would we consider:

$$\begin{array}{cc} 14 & -1 \\ 1 & +12 \end{array} \qquad\qquad \begin{array}{cc} 14 & +1 \\ 1 & -12 \end{array}$$

because AD is very large and BC is small. The difference must be far in excess of 13. Then:

$$14x^2 + 13x - 12 = (7x - 4)(2x + 3)$$

3. Factorize $15x^2 - 26x + 7$.
Assume $15x^2 - 26x + 7 = (A - B)(C - D)$
Possible values of A and C are $15x$, x; x, $15x$; $5x$, $3x$; $3x$, $5x$
Possible values of B and D are 1, 7; 7, 1
All values of A, B, C and D are odd. We must in all cases obtain an even value for $-AD - BC$. We do not even try:

$$\begin{array}{cc} 15 & -1 \\ 1 & -7 \end{array} \qquad\qquad \begin{array}{cc} 5 & -1 \\ 3 & -7 \end{array}$$

because they would produce a middle term too large numerically. We try:

$$\begin{array}{cc} 15 & -7 \\ 1 & -1 \end{array}$$ The middle term is $-22x$. Reject.

$$\begin{array}{cc} 5 & -7 \\ 3 & -1 \end{array}$$ The middle term is $-26x$. Accept.

Therefore $15x^2 - 26x + 7 = (5x - 7)(3x - 1)$

4. Factorize, if possible, $21x^2 + 37x + 25$.
This expression cannot factorize because all possible pairs of factors of 21 and 25 are odd, and they must produce a middle term which is even.

5. Factorize $72x^2 + 190x - 22$.
First: $72x^2 + 190x - 22 = 2(36x^2 + 95x - 11)$
Assume $36x^2 + 95x - 11 = (A + B)(C - D)$
Possible values of A and C: $36x$, x; x, $36x$; $18x$, $2x$; $2x$, $18x$; $9x$, $4x$; $4x$, $9x$; $3x$, $12x$; etc.
Possible values of B and D: 11, 1; 1, 11
Do not consider:

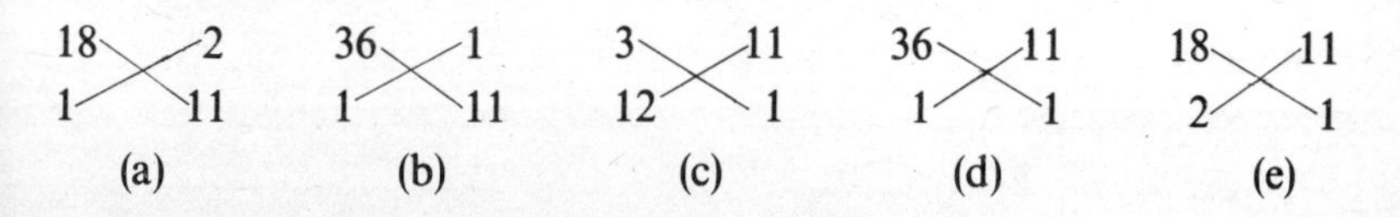

The middle term in (a) is even; in (b) and (c) it is too large; and in (d) and (e) it is too small. The correct combination is:

$$\begin{matrix} 9 & \searrow\!\!\!\!\nearrow & -1 \\ 4 & \nearrow\!\!\!\!\searrow & +11 \end{matrix}$$

Then:

$$72x^2 + 190x - 22 = 2\,(9x - 1)\,(4x + 11)$$

6. Apply Method 2 to the factorization of $14x^2 + 13x - 12$. The previous method can prove tedious when the correct combination is not obtained quickly. Method 2 would also be tedious if no calculator were available to determine squares and square roots of large numbers accurately. It does have the advantage that the outcome is certain.

$$14x^2 + 13x - 12 = 14\left(x^2 + \frac{13}{14}x - \frac{12}{14}\right)$$

$$= 14\left[x^2 + \frac{13}{14}x + \left(\frac{13}{28}\right)^2 - \frac{12}{14} - \frac{169}{784}\right]$$

In the bracket add the square of half the coefficient of x to complete the square:

$$14\left[\left(x + \frac{13}{28}\right)^2 - \frac{672}{784} - \frac{169}{784}\right] = 14\left[\left(x + \frac{13}{28}\right)^2 - \frac{841}{784}\right] \quad \text{using } \mathit{1.2}$$

$$= 14\left[\left(x + \frac{13}{28}\right)^2 - \left(\frac{29}{28}\right)^2\right]$$

$$= 14\left(x + \frac{13}{28} + \frac{29}{28}\right)\left(x + \frac{13}{28} - \frac{29}{28}\right) \quad \text{by } \mathit{1.4}$$

$$= 14\left(x + \frac{42}{28}\right)\left(x - \frac{16}{28}\right) = 14\left(x + \frac{3}{2}\right)\left(x - \frac{4}{7}\right)$$

$$= 2\left(x + \frac{3}{2}\right)7\left(x - \frac{4}{7}\right) = (2x + 3)\,(7x - 4)$$

Exercise 1.3

Factorize:

1. $2x^2 - 3x + 1$
2. $2x^2 + 3x + 1$

3. $2x^2+x-1$
4. $2x^2-x-1$
5. $3y^2+4y+1$
6. $3y^2-4y+1$
7. $3y^2+2y-1$
8. $3y^2-2y-1$
9. $4p^2+5p+1$
10. $4p^2-5p+1$
11. $4p^2+3p-1$
12. $4p^2-3p-1$
13. $2q^2+5q+2$
14. $2q^2-5q+2$
15. $2q^2+3q-2$
16. $2q^2-3q-2$
17. $3x^2+7x+2$
18. $3a^2-7a+2$
19. $3b^2+5b-2$
20. $3c^2-5c-2$
21. $6x^2-13x+6$
22. $6x^2+5x-6$
23. $5y^2+8y+3$
24. $5x^2-16x+3$
25. $7z^2+13z+6$
26. $7x^2-17x+6$
27. $7a^2+23a+6$
28. $7b^2+11b-6$
29. $7x^2-19x-6$
30. $12x^2+25x+12$
31. $12x^2+7x-12$
32. $12x^2-10x-12$

1.2 Non-factorizable quadratic expressions

Example (4) above is an expression which is non-factorizable. There are many more which cannot be factorized. Examples are: x^2+3x+1, $x^2-4x+1, x^2 \pm px+1$, where $p > 2$. A very special kind is x^2+1; more generally, x^2+y^2. There are no real factors for such an expression. Particular examples of this are: $x^2+4, x^2+49, 4x^2+1, 4a^2+25$, $121p^2q^2+64r^2$.

Exercise 1.4

Determine which of the following expressions do not factorize:

1. $x^2 - 4x + 3$
2. $x^2 - 23x + 3$
3. $a^2 - 11a + 10$
4. $p^2 - 9a + 10$
5. $x^2 + x + 24$
6. $a^2 + b^2$
7. $a^2 + (-b)^2$
8. $a^2 + (2b)^2$
9. $(3a)^2 + (-2b)^2$
10. $(x+1)^2 + (y+1)^2$
11. $(x+1)^2 + (y-1)^2$
12. $(x+1)^2 + (x-1)^2$

1.3 The roots of an equation

An equation is the statement that two expressions are equal. Examples are $2x + 5 = 0$; $3/y = 2$; $a^2 + a = 2$; $(x-2)(x-5) = 0$. They are satisfied for a limited number of values of the unknown.

Examples

1. Consider $2x + 5 = 0$.
   This leads to:

$$2x + 5 - 5 = -5, \quad \text{i.e.}$$
$$2x = -5 \quad \text{or} \quad x = -5/2$$

   $-5/2$ is said to be the root of the equation $2x + 5 = 0$. If we substitute $-5/2$ for x in the LHS of the equation we obtain:

$$2(-5/2) + 5 = -5 + 5 = 0, \quad \text{i.e. the RHS}$$

   The equation $2x + 5 = 0$ is said to be satisfied by $(-5/2)$, i.e. by the root
2. Consider $(x-2)(x-5) = 0$.
   If we put $x = 2$ in the LHS we obtain $(2-2)(2-5) = 0(-3) = 0$.
   If we put $x = 5$ in the LHS we obtain $(5-2)(5-5) = 3(0) = 0$.
   This means that 2 and 5 are roots of the equation.

3. Write down the equation with roots 1 and 2.
   By looking at (2) above we might guess the equation is $(x-1)(x-2) = 0$, because by substituting $x = 1$ and $x = 2$ separately the equation is satisfied.
4. Write down the equation with roots 3 and -2.
   Here we write:

$$(x-3)(x-{}^{-}2) = 0, \quad \text{i.e.}$$
$$(x-3)(x+2) = 0$$

Exercise 1.5

Write down the equations with the following roots:

1. 1, 3
2. 2, 3
3. 4, 5
4. 1, -1
5. 2, -1
6. 3, -2
7. 4, -5
8. a, $2a$
9. a, $-a$
10. $\frac{1}{2}$, 1

1.4 Equations with a given pair of roots

The answer to question (10) in Exercise 1.5 would appear to be:

$$(x-\tfrac{1}{2})(x-1) = 0$$

That is the answer. But a better answer is:

$$2(x-\tfrac{1}{2})(x-1) = 0$$

since $2 \times 0 = 0$, which gives:

$$(2x-1)(x-1) = 0$$

That is an equation without fractional coefficients. In general an equation with roots p and q is:

$$(x-p)(x-q) = 0 \qquad 1.5$$

By expanding the RHS we obtain:

$$x^2 - px - qx + pq = 0$$

$$\text{or} \quad x^2 - (p+q)x + pq = 0 \qquad 1.6$$

i.e. $$x^2 - (\text{sum of roots})\,x + (\text{product of roots}) = 0 \qquad 1.7$$

1.5, *1.6* and *1.7* are alternative forms of equations with roots p and q.

Exercise 1.6

Write down (a) in factor form, (b) in factor form without fractional coefficients, and (c) in expanded form, the equations with the following pairs of roots:

1. $1, \frac{1}{4}$
2. $\frac{1}{2}, \frac{1}{3}$
3. $\frac{1}{2}, -1$
4. $\frac{1}{2}, -\frac{1}{3}$
5. $1\frac{1}{2}, -3\frac{1}{2}$
6. $-4\frac{1}{2}, -2\frac{1}{2}$
7. $-3\frac{1}{2}, -2\frac{1}{4}$
8. $0, 1$
9. $-2\frac{1}{2}, 0$
10. a, b
11. $a, -a$
12. $-a, -a$
13. $a/b, 2a/b$
14. $a/b, b/a$
15. $a, 1/a$
16. $\dfrac{a+b}{c}, \dfrac{a-b}{c}$

1.5 Quadratic expressions and equations

The basic form of a quadratic expression is:

$$ax^2 + bx + c, \text{ where } a, b \text{ and } c \text{ are constants and } a \neq 0 \qquad 1.8$$

The basic form of a quadratic equation is:

$$px^2 + qx + r = 0, \text{ where } p, q \text{ and } r \text{ are constants and } p \neq 0 \qquad 1.9$$

Examples

1. $3(2x^2 - 5x - 3) + 4(5 - 4x + 2x^2)$
   This can be written:

$$6x^2 - 15x - 9 + 20 - 16x + 8x^2 = 14x^2 - 31x + 11$$

   which is a quadratic expression.
2. $(3x + 2)(4x - 5) + 2(2x - 3) + 7$
   This can be written:

$$12x^2 + 8x - 15x - 10 + 4x - 6 + 7 = 12x^2 - 3x - 9$$

   which is a quadratic expression.
3. $5x^2 - 12x - 4 = 0$ is a quadratic equation.
4. $13x^2 = 6x + 2$
   This leads to:

$$13x^2 - 6x - 2 = 6x + 2 - 6x - 2, \text{ i.e.}$$
$$13x^2 - 6x - 2 = 0$$

   which is a quadratic equation.
5. $3(x + 2) = 5(1/x - 4)$ leads to:

$$3x + 6 = 5/x - 20$$

   Multiply both sides by x:

$$3x^2 + 6x = 5 - 20x$$

   giving:

$$3x^2 + 26x - 5 = 0$$

   which is a quadratic equation.
6. $2(\frac{1}{x} - 5) = \frac{1}{x}(3 - 4x)$ leads to:

$$2/x - 10 = 3/x - 4$$

   Multiply both sides by x:

$$2 - 10x = 3 - 4x, \text{ i. e.}$$
$$0 = 6x + 1$$

   which is not a quadratic equation.

Exercise 1.7

Determine which of the following are quadratic expressions and which are quadratic equations:

1. $3x^2+2x+1$
2. $(x-1)(x-2)+3(x+1)+2$
3. $2x(4x+1/x)+6x+2$
4. $3x^2+1$
5. $3x^2+2x$
6. $3x^2+2x+1=0$
7. $3x^2+1=0$
8. $3x^2-2x=0$
9. $2x(4x+3-2/x)-4x(2x+1-3/x)$
10. $x(x+1)(2x+1)-2x^3=0$
11. $x(x+1)(2x+1)+2x^3$
12. $\dfrac{x^2+2x+3}{2x+1}=2$
13. $\dfrac{x-1}{x+1}+\dfrac{2x+1}{3x-1}=0$
14. $\dfrac{x-1}{x+1}+\dfrac{2x+1}{3x+1}$

1.6 The solution of quadratic equations

Solution by factorization

In this method the basic principle is the following. Where $p \times q = 0$ and p and q represent numbers which may take any value, there are three possible conclusions:

$$\left.\begin{array}{ll}\text{(a)} & p \text{ must equal zero} \\ \text{(b)} & q \text{ must equal zero} \\ \text{(c)} & \text{both } p \text{ and } q \text{ must equal zero}\end{array}\right\} \quad 1.10$$

When we apply this to a quadratic equation in the form

$$(x-1)(x-2)=0$$

we think of replacing $x-1$ by p and $x-2$ by q. Then:

$$\begin{array}{rl}\text{either (a)} & x-1=0, \text{ i.e. } x=1, \\ \text{or (b)} & x-2=0, \text{ i.e. } x=2\end{array}$$

In this case (c) does not apply because x cannot equal 1 and 2 at the same time. Therefore either $x=1$ or $x=2$. They are the solutions of the equation. If we refer back to section 1.4, another way of expressing this is to say that 1 and 2 are the roots of the equation.

Examples

1. Solve $(3x-1)(2x+5)=0$.
   Then

$$3x-1=0 \quad \text{or} \quad 2x+5=0$$
$$3x=1 \quad \text{or} \quad 2x=-5$$
$$x=\tfrac{1}{3} \quad \text{or} \quad x=-2\tfrac{1}{2}$$

   Check $x=\frac{1}{3}$: LHS $=(1-1)(\frac{2}{3}+5)=0\times 5\frac{2}{3}=0$
   Check $x=-2\frac{1}{2}$: LHS $=(-7\frac{1}{2}-1)(-5+5)=-8\frac{1}{2}\times 0=0$
   The roots are $\frac{1}{3}$ and $-2\frac{1}{2}$.
2. Solve $4s^2-s+3=2s(s-4)$.
   Rearrange:

$$4s^2-s+3=2s^2-8s$$

   Leading to:

$$2s^2+7s+3=0$$

   Factorize the LHS: $(2s+1)(s+3)=0$. Then:

$$2s+1=0 \quad \text{or} \quad s+3=0$$
$$s=-\tfrac{1}{2} \quad \text{or} \quad s=-3$$

   Check $s=-\frac{1}{2}$: LHS $=4(\frac{1}{4})+\frac{1}{2}+3=4\frac{1}{2}$; RHS $=-1(-\frac{1}{2}-4)=4\frac{1}{2}$
   Check $s=-3$: LHS $=36+3+3=42$; RHS $=-6(-3-4)=42$
   The roots are $-\frac{1}{2}$ and -3.
3. Solve $(3x-2)(4x+5)+10=0$.
   Then:

$$12x^2+7x-10+10=0$$
$$12x^2+7x=0$$
$$x(12x+7)=0$$
$$x=0 \quad \text{or} \quad 12x+7=0$$
$$x=0 \quad \text{or} \quad x=-7/12$$

   Check $x=0$: LHS $=(0-2)(0+5)+10=-10+10=0$
   Check $x=-7/12$: LHS $=(-7/4-2)(-7/3+5)+10$
   $=-3\frac{3}{4}\times 2\frac{2}{3}+10=-\dfrac{15}{4}\times\dfrac{8}{3}+10=-10+10=0$
   The roots are 0 and $-7/12$.

4. Solve $(2x+3)(2x+5) = 4(4x+6)$.
   Then:

$$4x^2 + 16x + 15 = 16x + 24$$
$$4x^2 - 9 = 0$$
$$(2x+3)(2x-3) = 0$$
$$2x+3 = 0 \quad \text{or} \quad 2x-3 = 0$$
$$x = -3/2 \text{ or } x = 3/2$$

   Check $x = 3/2$: LHS $= (3+3)(3+5) = 6 \times 8 = 48$; RHS $= 4(6+6) = 48$
   Check $x = -3/2$: LHS $= (-3+3)(-3+5) = 0$; RHS $= 4(-6+6) = 0$
   The roots are $3/2$ and $-3/2$.

Exercise 1.8

Solve the following equations:

1. $(2x+5)(3x-2) = 0$
2. $(4x-1)(x+4) = 0$
3. $x(x-2) = 0$
4. $2x(x-2) = 0$
5. $3x(5x-2) = 0$
6. $(2d-1)(3d-4) = 0$
7. $s^2 + 3s + 2 = 0$
8. $4p^2 - p - 3 = 0$
9. $(3t-1)(2t-1) = 26$
10. $\dfrac{3x+1}{13} = \dfrac{2}{2x+1}$

The solution of quadratic equations by completing the square and the formula

The completing the square method leads to a formula for the solution. The following is the basic principle:

$$x^2 = a^2 \text{ has roots}$$
$$x = \pm a$$

by taking the square root of the equation above. This can be extended to:

$$(x-p)^2 = q^2 \text{ has roots given by:}$$
$$x - p = \pm q \text{, i.e. } x = p \pm q$$

1.11

Examples

1. Solve $x^2+2x-5=0$.
   Write:

$$x^2+2x=5$$
$$x^2+2x+1=5+1=6$$

The LHS is now a perfect square, i.e.

$$(x+1)^2=6$$

Then

$$x+1=\pm\sqrt{6}$$

The roots are $x=-1\pm\sqrt{6}$.

2. Solve the equation $ax^2+bx+c=0$.
   Divide by a:

$$x^2+\frac{b}{a}x+\frac{c}{a}=0$$
$$x^2+\frac{b}{a}x=-\frac{c}{a}$$

Complete the square of the LHS:

$$x^2+\frac{b}{a}x+\left(\frac{b}{2a}\right)^2=\left(\frac{b}{2a}\right)^2-\frac{c}{a}$$
$$\left(x+\frac{b}{2a}\right)^2=\frac{b^2}{4a^2}-\frac{4ac}{4a^2}=\frac{b^2-4ac}{4a^2}$$
$$x+\frac{b}{2a}=\pm\frac{\sqrt{b^2-4ac}}{\sqrt{4a^2}}=\pm\frac{\sqrt{b^2-4ac}}{2a}$$
$$x=\frac{-b\pm\sqrt{b^2-4ac}}{2a} \qquad 1.12$$

These are the roots of the equation.

This is also the formula which we use to solve those quadratic equations which we cannot solve by the factor method.

Examples

1. Use the formula method to solve $3x^2+5x-7=0$. (i)
   The general quadratic equation is $ax^2+bx+c=0$. In (i) $a=3$, $b=5$

and $c = -7$. Substitute in *1.12* for a, b and c:

$$x = \frac{-5 \pm \sqrt{25 - 4.3(-7)}}{6}$$

$$= \frac{-5 \pm \sqrt{25+84}}{6} = \frac{-5 \pm \sqrt{109}}{6} = \frac{-5 \pm 10.440307}{6} \text{ by calculator}$$

$$= \frac{-5 + 10.440307}{6} \text{ or } \frac{-5 - 10.440307}{6} = +\frac{5.440307}{6} \text{ or } -\frac{15.440307}{6}$$

$$= 0.9067178 \text{ or } -2.5733845 \approx 0.907 \text{ or } -2.573$$

Check $x = 0.906718$: LHS $= 3(0.906718)^2 + 5(0.906718) - 7$
$= 2.4664126 + 4.53359 - 7 = 0.0000026.$
Check $x = -2.5733845$: LHS $= 3(-2.5733845)^2 + 5(-2.5733845) - 7$
$= 19.866923 - 12.866923 - 7 = 0.0000009$

2. Solve $2x^2 - 8x + 5 = 0$. (ii)
The general quadratic equation is $ax^2 + bx + c = 0$. In (ii) $a = 2$, $b = -8$ and $c = 5$. Substitute in *1.12* for a, b and c. The roots are:

$$\frac{8 \pm \sqrt{(-8)^2 - 4.2.5}}{4}$$

$$= \frac{8 \pm \sqrt{64 - 40}}{4}$$

$$= \frac{8 \pm 4.8989795}{4}$$

$$= \frac{12.8989795}{4} \text{ or } \frac{3.1010205}{4}$$

$$= 3.2247449 \text{ or } 0.7752551$$

$$\approx 3.225 \text{ or } 0.775$$

Now check these answers as in example 1.

Exercise 1.9

Solve the following equations by the formula:

1. $2x^2 + 3x - 4 = 0$
2. $4x^2 + 4x - 1 = 0$
3. $2x^2 + 6x - 5 = 0$
4. $2x^2 + 3x - 4 = 0$

5. $2x^2 - 6x - 5 = 0$
6. $-5 - 6x + 2x^2 = 0$
7. $5 - 6x - 2x^2 = 0$
8. $2x^2 + 10x + 3 = 0$
9. $2x + 10 + 3/x = 0$
10. $10x - 3 - 4x^2 = 0$
11. $2x - 3 - 2/x = 0$
12. $p(p-2) = 6$
13. $a^2 - 4a = 1$
14. $3x^2 + 5x - 4 = 0$

1.7 Applications of quadratic equations

Examples

1. A ball is thrown vertically upwards with an initial velocity of 30 m/s. Determine the time taken to reach a height of 35 m. Take the acceleration due to gravity to be 9.8 m/s^2. Fig. 1.1 is a diagram representing the motion.

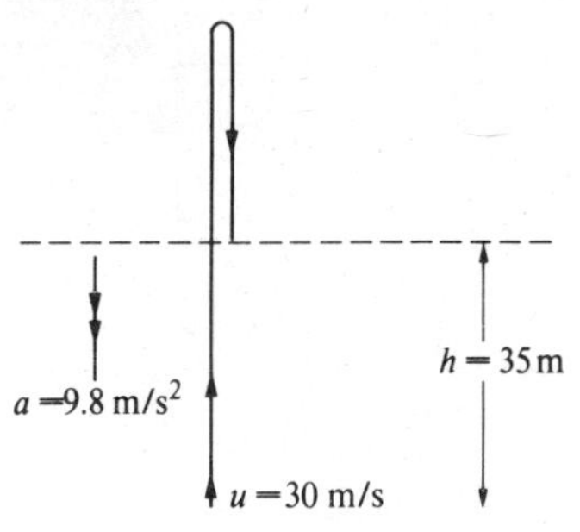

Figure 1.1

Suppose the time taken to reach height 35 m above the point of projection is t seconds. Then, from the formula for uniform acceleration, $h = ut - \frac{1}{2}at^2$, we obtain:

$$35 = 30t - \tfrac{1}{2}9.8t^2$$
$$35 = 30t - 4.9t^2, \text{ or}$$
$$4.9t^2 - 30t + 35 = 0$$

Here $a = 4.9$, $b = -30$ and $c = 35$. By substituting in the formula:

$$t = \frac{30 \pm \sqrt{(-30)^2 - 4 \times 4.9 \times 35}}{9.8}$$

$$= \frac{30 \pm \sqrt{900 - 686}}{9.8} = \frac{30 \pm \sqrt{214}}{9.8} = \frac{30 \pm 14.628739}{9.8}$$

$$= \frac{30 + 14.628739}{9.8} \text{ or } \frac{30 - 14.628739}{9.8}$$

$$= 4.5539529 \text{ or } 1.568496 \approx 4.55 \text{ or } 1.57$$

Note the two answers arising from the quadratic equation. The smaller answer refers to the time taken to pass the level 35 m as the ball moves upwards. The larger answer refers to the time taken to pass that level as the ball later moves downwards. If we were interested only in the former value we would reject the value 4.55.

2. Fig. 1.2 represents an electrical circuit containing three resistors, two of which are known to have equal resistances but are of unknown value. Determine the value of the unknown resistances if the total resistance of the circuit is 30 Ω.

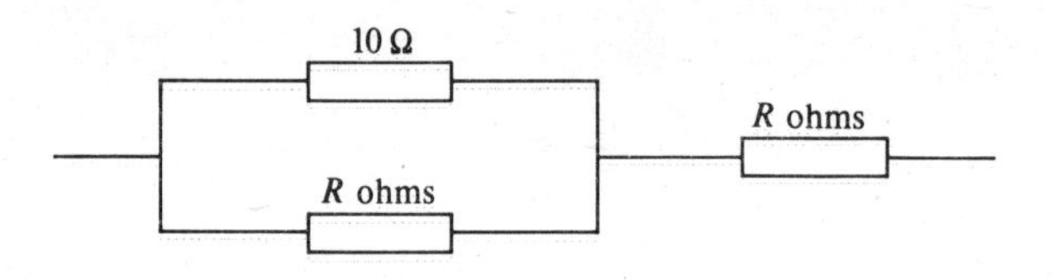

Figure 1.2

Suppose the equivalent resistance of the parallel circuit is R'. Then:

$$1/R' = 1/10 + 1/R = \frac{R + 10}{10R}$$

giving:

$$R' = \frac{10R}{R + 10}$$

Then the total resistance $= R' + R = 30$. Therefore:

$$\frac{10R}{R + 10} + R = 30$$

Multiply throughout by $R + 10$:

$$10R + R(R+10) = 30(R+10)$$
$$10R + R^2 + 10R = 30R + 300$$
$$R^2 - 10R - 300 = 0$$

The LHS will not factorize. Put $a = 1$, $b = -10$ and $c = -300$ in the formula, to give:

$$R = \frac{10 \pm \sqrt{(-10)^2 - 4 \times 1 \times (-300)}}{2}$$
$$= \frac{10 \pm \sqrt{100 + 1200}}{2} = \frac{10 \pm \sqrt{1300}}{2}$$
$$R = \frac{10 + \sqrt{1300}}{2} \text{ or } \frac{10 - \sqrt{1300}}{2} = 23.027756 \text{ or } -13.027756$$

Then $R \approx 23.03$, since we reject the negative answer. The two resistors have a resistance of approximately 23.03 Ω.
Check: $23.027756^2 - 10 \times 23.027756 - 300$
$= 530.27754 - 230.27756 - 300 = -0.00002$

3. In a chemical reaction between 3 mol of acetic acid and 5 mol of ethanol during which ethyl acetate is produced, suppose that x mole of ester and water are produced at equilibrium. Determine x.
The equilibrium law gives:

$$\frac{x^2}{(3-x)(5-x)} = 4$$

from which we obtain:

$$3x^2 - 32x + 60 = 0$$

From the formula:

$$x = \frac{32 \pm \sqrt{(-32)^2 - 4 \times 3 \times 60}}{6}$$
$$= \frac{32 \pm \sqrt{1024 - 720}}{6} = \frac{32 \pm \sqrt{304}}{6}$$
$$= \frac{32 \pm 17.435596}{6}$$
$$= \frac{32 + 17.435596}{6} \text{ or } \frac{32 - 17.435596}{6}$$
$$= 8.239266 \text{ or } 2.4274007 \approx 8.239 \text{ or } 2.43 \text{ mol.}$$

We reject the answer 8.239 because that is greater than the total amount of acetic acid and ethanol in the first place. Therefore approximately 2.43 mol each of ester and water are formed at equilibrium.

4. A rectangular room is 4 m longer than it is high and 2 m longer than it is wide. The total wall surface is $54\ \text{m}^2$. Determine the height of the room. Suppose the height of the room is x metres. Then:

$$2x(x+4)+2x(x+2) = 54$$
$$4x^2+12x = 54$$
$$x^2+3x-18 = 0$$
$$(x+6)(x-3) = 0$$
$$\text{Therefore } x = 3 \text{ or } x = -6$$

We reject the answer -6. The height of the room is 3 m.

Exercise 1.10

1. A ball is thrown vertically upwards at 40 m/s. Determine the time taken for it to reach a height 50 m above the point of projection. Take the acceleration due to gravity to be $9.8\ \text{m/s}^2$.
2. Two unknown equal resistances are in series in a circuit. A third resistor, of 15 Ω resistance, is connected in parallel with one of the first two. The total resistance of that circuit is 32 Ω. Determine the resistance of the two unknown resistors.
3. The circuit represented by Fig. 1.3 has a total resistance of 40 Ω. Determine the value of R.

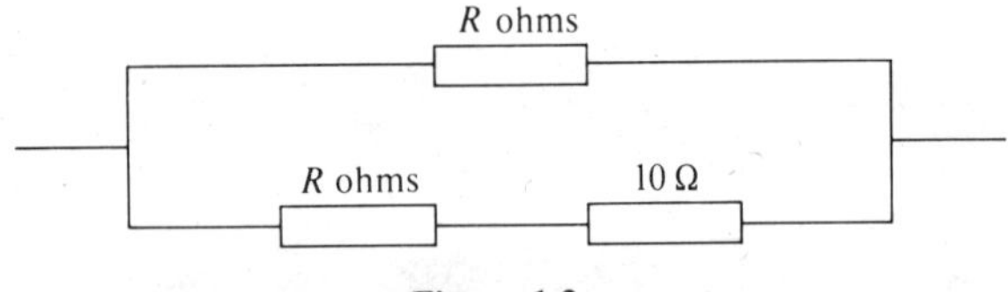

Figure 1.3

4. The circuit represented by Fig. 1.4 has a total resistance of 25 Ω. Determine the value of R.

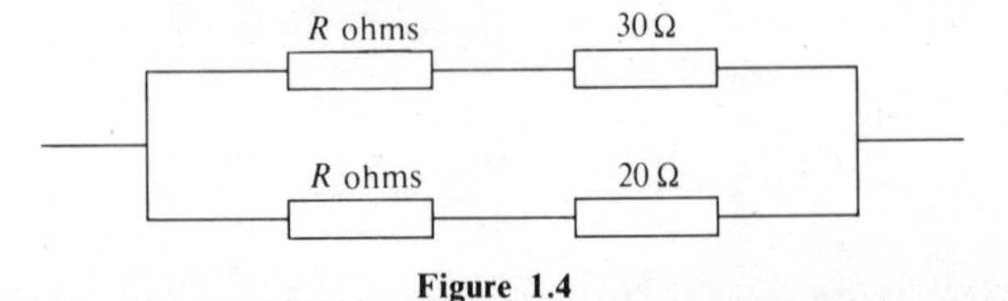

Figure 1.4

5. A completely enclosed hollow body has the shape illustrated in Fig. 1.5. Its total surface area is given by $S = \pi r^2 + 2\pi rh + 2\pi r^2$. Determine r given that $S = 30\pi$ and $h = 4.3$.

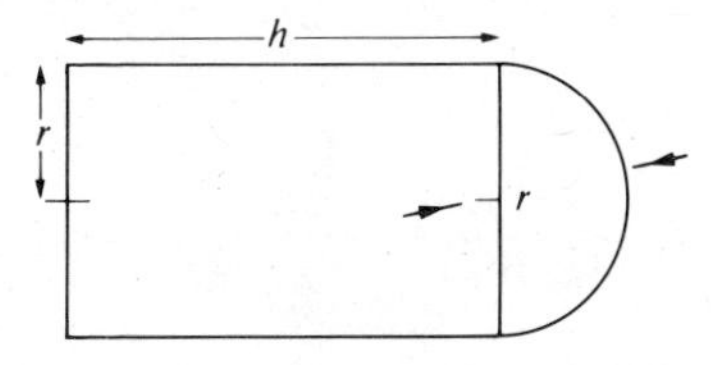

Figure 1.5

6. The equation $Lm^2 + Rm - 1/C = 0$ arises in certain electrical circuits. Determine m given that $L = 1.53$, $C = 10.3 \times 10^{-6}$ and $R = 8.7 \times 10^3$.
7. A rectangular room is 6 m longer than it is high and 3.7 m longer than it is wide. The total wall surface is 70 m^2. Determine the height of the room.
8. Find the width of a uniform border which is to be placed round a rectangular area 10.3 m × 14.7 m, if the border area is to be 53.8 m^2.
9. Fig. 1.6 represents a circuit containing a resistance, R, in series with an unknown unit. If the voltage drop across the unknown unit is V then the power consumption of the circuit is $I^2R + IV$. Determine I when the power consumption is 16 120 W, the voltage drop is 410 V and the resistance, R, is 11.3 Ω.

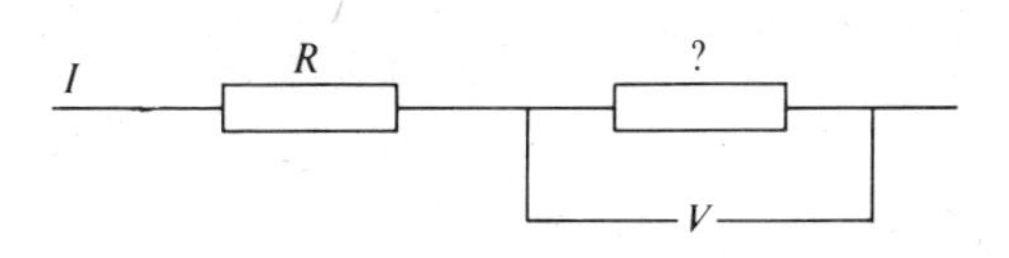

Figure 1.6

10. Assuming the equilibrium law when 1 mol of hydrogen and 2.1 mol of iodine react to equilibrium to be:

$$\frac{(1-x)(2.1-x)}{(2x)^2} = \frac{1}{64}$$

determine the value of x.

1.8 The solution of simultaneous equations (one linear, one quadratic)

Examples

1. Solve:

$$x^2 + y^2 = 25 \qquad \text{(i)}$$

$$4y = 3x \qquad \text{(ii)}$$

From (ii) $$y = \frac{3}{4}x \qquad \text{(iii)}$$

Substitute for y from (iii) in (i):

$$x^2 + \frac{9}{16}x^2 = 25$$

$$\frac{25}{16}x^2 = 25$$

$$x^2 = 25 \times \frac{16}{25} = 16$$

$$x = \pm 4$$

Substitute for x in (iii):

$$y = \pm\frac{3}{4} \times 4 = \pm 3$$

The solutions are: (a) $x = 4$, $y = 3$; (b) $x = -4$, $y = -3$.
Check (a): Substitute in (i), LHS $= 16 + 9 = 25$
Substitute in (ii), LHS $= 4 \times 3 = 12$, RHS $= 3 \times 4 = 12$
Check (b): Substitute in (i), LHS $= 16 + 9 = 25$
Substitute in (ii), LHS $= 4 \times -3 = -12$, RHS $= 3 \times -4 = -12$

Note that we always substitute in the original equations because, in writing down any subsequent equation, error might have been introduced.

2. Solve:

$$xy = 3 \qquad \text{(i)}$$

$$x - 2y = 5 \qquad \text{(ii)}$$

From (ii) $$x = 2y + 5 \qquad \text{(iii)}$$

Substitute for x in (i):

$$y(2y + 5) = 3$$

$$2y^2 + 5y - 3 = 0$$

$$(2y - 1)(y + 3) = 0$$

$$y = \tfrac{1}{2} \text{ or } -3$$

Substitute in (iii):

$$x = 1 + 5 = 6, \text{ or } -6 + 5 = -1$$

The solutions are: (a) $x = 6$, $y = \frac{1}{2}$; (b) $x = -1$, $y = -3$.
Check (a): Substitute in (i), LHS $= 6 \times \frac{1}{2} = 3$
Substitute in (ii), LHS $= 6 - 1 = 5$
Check (b): Substitute in (i), LHS $= -1 \times -3 = 3$
Substitute in (ii), LHS $= -1 - (2 \times -3) = -1 + 6 = 5$

3. Solve:

$$x^2 + 2xy - y^2 + 28 = 0 \qquad \text{(i)}$$
$$3x + y = 2 \qquad \text{(ii)}$$

From (ii) $\quad y = 2 - 3x \qquad$ (iii)

Substitute for y from (iii) in (i):

$$x^2 + 2x(2 - 3x) - (2 - 3x)^2 + 28 = 0$$
$$x^2 + 4x - 6x^2 - 4 + 12x - 9x^2 + 28 = 0$$
$$-14x^2 + 16x + 24 = 0 \qquad \text{(iv)}$$

(iv) $\div (-2)$
$$7x^2 - 8x - 12 = 0$$
$$(7x + 6)(x - 2) = 0$$
$$x = 2 \text{ or } -6/7$$

Substitute for x in (iii):

$$y = -4 \text{ or } 4\tfrac{4}{7}$$

The solutions are: (a) $x = 2$, $y = -4$; (b) $x = -6/7$, $y = 4\frac{4}{7}$.
Check (a): Substitute in (i), LHS $= 4 - 16 - 16 + 28 = 0$
Substitute in (ii), LHS $= 6 - 4 = 2$
Check (b): Substitute in (i), LHS $= \frac{36}{49} - \frac{384}{49} - \frac{1024}{49} + 28$

$$= -\frac{1372}{49} + 28 = -28 + 28 = 0$$

Substitute in (ii), LHS $= -\frac{18}{7} + \frac{32}{7} = \frac{14}{7} = 2$

Exercise 1.11

Solve the following:

1. $x + y = 8$, $x^2 + y^2 = 34$
2. $x - y = 2$, $xy = 24$

3. $2x - y = 1, 4x^2 + y^2 = 61$
4. $x - y = 11, xy + 28 = 0$
5. $3x + 2y = -1, x^2 - 4xy + 3y^2 = 21$
6. $2x - y = 2, xy = 12$
7. $4x - 5y = 0, 2x^2 - xy - y^2 = 14$
8. $2x + y + 1 = 0, 12x^2 - xy - y^2 = .14$
9. $3x + 2y = 14, 3x^2 + 2xy + 4y^2 = 60$
10. $x + y = 3, x^2 - y^2 = 12$
11. $x - y = 3, x^2 - y^2 = 12$
12. $x - 2y = 5, x^2 - 4y^2 = 35$
13. $2x - 5y = 11, 6x^2 - 11xy - 10y^2 = 165$
14. $7x + 4y = 29, 14x^2 - 13xy - 12y^2 = 0$
15. $2x - y - 3 = 0, x^2 + y^2 = 26$

2

Application of the Laws of Logarithms

2.1 Logarithms to any base

In earlier work at Level 1 we discovered that any positive number, N, may be written in the form 10^x. To every such number, N, there is an index, x. They form a one-to-one relationship. Each value of N determines just one value of x and each value of x determines just one value of N. There are two different ways of expressing this relationship:

$$N = 10^x \qquad 2.1$$

$$x = \log N \qquad 2.2$$

2.1 makes N the subject of the relationship, while *2.2* makes x the subject of the relationship. In *2.1* and *2.2*, 10 is said to be the base of the logarithms. There we are dealing with logarithms to base 10, often called common logarithms.

But a number-index relationship, i.e. a number-logarithm relationship, can be based on any real positive number. For example, one number used as a base in much of technology and science, as well as in statistics, is e, which is incommensurable, and has an approximate value of 2.7182818285. We write:

$$N = \mathrm{e}^p \qquad 2.3$$

$$\text{or } p = \log_{\mathrm{e}} N, \text{ or } \ln N \qquad 2.4$$

Suppose, however, we choose any real positive number, b, as a base. Then:

$$N = b^y \qquad 2.5$$

Leading to:

$$y = \log_b N \qquad 2.6$$

2.5 makes N the subject and *2.6* makes y the subject of the relationship. This means that the two functions represented in *2.5* and *2.6* are inverses of each other. Substitute from *2.6* for y in *2.5* to obtain:

$$N = b^{\log_b N} \qquad 2.7$$

The RHS of *2.7* is a pair of functions operating on N in succession. They are inverses because the result is N, on the left, and that means the whole operation on the right leaves N unchanged. By substituting from *2.5* for N in *2.6* we obtain:

$$y = \log_b b^y \qquad 2.8$$

This time the operation on y on the right leaves y unchanged. The basic laws which apply to logarithms to base 10 also apply to logarithms to any other base.

$$\text{Suppose } M = b^x, \quad \text{then} \quad x = \log_b N \qquad 2.9$$

$$\text{Suppose } N = b^y, \quad \text{then} \quad y = \log_b N \qquad 2.10$$

$$\text{This means} \quad M \times N = b^x \times b^y = b^{x+y} \qquad 2.11$$

$$\text{Therefore} \quad x + y = \log_b (M \times N) \qquad 2.12$$

Substituting from *2.9* and *2.10* in *2.12*:

$$\log_b (M \times N) = \log_b M + \log_b N \qquad 2.13$$

Again, $M/N = b^x/b^y = b^{x-y}$. Therefore:

$$x - y = \log_b (M/N) \qquad 2.14$$

Substituting from *2.9* and *2.10* in *2.14*:

$$\log_b (M/N) = \log_b M - \log_b N \qquad 2.15$$

Further:

$$M^n = (b^x)^n = b^{nx} \qquad 2.16$$

This means:

$$\log_b (M^n) = nx \qquad 2.17$$

Substitute from *2.9*:

$$\log_b (M^n) = n \log_b M \qquad 2.18$$

2.13, *2.15* and *2.18* constitute the basic laws of logarithms related especially to logarithms to base b.

Certain special cases are important. From *2.9*, suppose $M = 1$. Here we know that $1 = a^0$, where a is any number. Therefore $b^0 = 1$. *2.9* gives:

$$b^0 = 1 = b^x, \text{ i.e. } x = 0$$

In other words:

$$\log_b 1 = 0 \qquad 2.19$$

If $M = b$, $M = b = b^1$. Therefore, in *2.9*

$$x = 1$$

$$\text{i.e. } \log_b b = 1 \qquad 2.20$$

If $M = 0$ then there is no value of x in *2.9* which fits. We can say that, as $M \to 0$, then $\log_b M \to -\infty$.

2.2 The relationship between indicial and logarithmic expressions

In certain cases the solution of indicial equations is easy.

Examples

1. Solve x when $a^x = a^p$
   Then $x = p$
2. Solve $3^x = 9^5$
   Then $3^x = (3^2)^5 = 3^{10}$
   Therefore $x = 10$
3. Solve $8^{x-3} = 32^{2x+1}$
   Then $(2^3)^{x-3} = (2^5)^{2x+1}$
   Therefore $2^{3x-9} = 2^{10x+5}$
   That is $3x - 9 = 10x + 5$
   Leading to $-14 = 7x$, $x = -2$

In other words, where the bases of the numbers on the two sides of the equation are easily related to one another, the method is clear. We change the base of one side into that of the other, or the two bases into a common base.

In the following examples the relationship between the two bases is not so simple and we have to modify the methods above.

Examples

1. Convert the following into its logarithmic equivalent: $16 = 4^2$.
   Take logarithms to base b of both sides:

$$\log_b 16 = \log_b 4^2 = 2\log_b 4 \qquad 2.21$$

Here the base of logarithms (b) can be of any value we choose so long as it is positive. Suppose we choose $b = 10$ (common logarithms), then *2.21* becomes:

$$\log 16 = 2\log 4 \qquad 2.22$$

By choosing $b = \mathrm{e}$ (natural logarithms), *2.21* becomes:

$$\ln 16 = 2\ln 4 \qquad 2.23$$

By selecting $b = 4$ and $b = 16$ respectively we obtain:

$$\log_4 16 = 2\log_4 4 = 2\times 1 = 2 \qquad 2.24$$

$$\text{and}\quad \log_{16} 16 = 2\log_{16} 4$$

$$\text{i.e.}\quad 1 = 2\log_{16} 4$$

$$\text{or}\quad \tfrac{1}{2} = \log_{16} 4 \qquad 2.25$$

2. Convert $243 = 3^5$ into its logarithmic equivalent.
   Taking logs to base b we obtain $\log_b 243 = \log_b 3^5 = 5\log_b 3$ *2.26*
   Taking logs to base 10: $\log 243 = 5\log 3$ *2.27*
   Taking logs to base e: $\ln 243 = 5\ln 3$ *2.28*
   Taking logs to base 243: $1 = 5\log_{243} 3$
   or $\log_{243} 3 = 1/5$ *2.29*
   Taking logs to base 3: $\log_3 243 = 5\log_3 3 = 5$ *2.30*
   Taking logs to base 9: $\log_9 243 = 5\log_9 3$ *2.31*
3. Solve the equation $4^x = 7^3$.
   Take logs to base 10:

$$\log 4^x - \log 7^3$$

$$x\log 4 = 3\log 7$$

$$x = \frac{3\log 7}{\log 4}$$

$$= \frac{3\times 0.845098}{0.60206} = \frac{2.5352941}{0.60206}$$

$$= 4.2110324 \approx 4.211$$

Check: substitute back in the original equation (by calculator):

$$\text{LHS} = 4^{4.2110324} = 343$$

$$\text{RHS} = 7^3 = 343$$

4. Solve the equation $11^{2x+9} = 7^{4x-3}$.
Take common logarithms:

$$\begin{aligned}(2x+9)\log 11 &= (4x-3)\log 7\\ 2x\log 11 + 9\log 11 &= 4x\log 7 - 3\log 7\\ 9\log 11 + 3\log 7 &= 4x\log 7 - 2x\log 11\\ (9\log 11 + 3\log 7) &= 2x\,(2\log 7 - \log 11)\end{aligned}$$

$$x = \frac{(9\log 11 + 3\log 7)}{2\,(2\log 7 - \log 11)} = \frac{9\times 1.0413927 + 3\times 0.845098}{2\,(2\times 0.845098 - 1.0413927)}$$

$$= \frac{9.3725342 + 2.5352941}{2\,(1.6901961 - 1.0413927)} = \frac{11.907828}{2\times 0.6488034}$$

$$= \frac{11.907828}{1.2976068} = 9.1767619 \approx 9.177$$

Check: substitute back in the original equation:

$$\text{LHS} = 3.060269\times 10^{28} \quad \text{RHS} = 3.0602708\times 10^{28}$$

5. The law $pv^{\gamma} = k$ represents the relationship between the pressure, p, and the volume, v, of a gas which expands adiabatically; γ is the ratio of the specific heats of the gas and k is a constant. Suppose, in fact, that the pressure and volume of a given mass of gas are p_1 and v_1 respectively in state 1 and that p_2 and v_2 are the corresponding values in state 2. Obtain the expression to determine the value of γ.
In state 1:

$$p_1 \,.\, v_1^{\gamma} = k$$

In state 2:

$$p_2 \,.\, v_2^{\gamma} = k$$

Then:

$$p_1 \,.\, v_1^{\gamma} = p_2 \,.\, v_2^{\gamma} \qquad \text{(i)}$$

Take common logarithms of (i):

$$\begin{aligned}\log p_1 + \log v_1^{\gamma} &= \log p_2 + \log v_2^{\gamma}\\ \log p_1 + \gamma\log v_1 &= \log p_2 + \gamma\log v_2\\ \gamma\log v_1 - \gamma\log v_2 &= (\log p_2 - \log p_1)\\ \gamma\,(\log v_1 - \log v_2) &= (\log p_2 - \log p_1)\\ \gamma\log(v_1/v_2) &= \log(p_2/p_1)\\ \gamma &= \frac{\log(p_2/p_1)}{\log(v_1/v_2)}\end{aligned}$$

Alternatively, from (i):

$$p_2/p_1 = v_1^{\gamma}/v_2^{\gamma} = (v_1/v_2)^{\gamma}$$

Now take logs of both sides:

$$\log(p_2/p_1) = \gamma \log(v_1/v_2)$$

giving:

$$\gamma = \frac{\log(p_2/p_1)}{\log(v_1/v_2)}$$

6. The law for a particular machine is of the form $P = aW^n$ where P and W represent the input and the output and a and n are constants. Determine an expression to evaluate n when P_1, W_1 and P_2, W_2 represent pairs of values of P and W. Then calculate n when $P_1 = 53.25$, $P_2 = 67.49$, $W_1 = 395.6$ and $W_2 = 503.8$.

$$P_1 = aW_1^n \qquad \text{(i)}$$

$$P_2 = aW_2^n \qquad \text{(ii)}$$

(i) ÷ (ii) gives:

$$\left(\frac{P_1}{P_2}\right) = \frac{W_1^n}{W_2^n} = \left(\frac{W_1}{W_2}\right)^n \qquad \text{(iii)}$$

Take logs of both sides of (iii):

$$n \log(W_1/W_2) = \log(P_1/P_2)$$

$$n = \frac{\log(P_1/P_2)}{\log(W_1/W_2)}$$

For the given data:

$$n = \frac{\log(53.25/67.49)}{\log(395.6/503.8)}$$

$$= \frac{\log(0.7890058)}{\log(0.7852322)} = \frac{-0.1029198}{-0.1050019} = 0.9801712 \approx 0.980$$

Check in (iii): LHS = 0.7890058; RHS = 0.7890058.

7. Simplify $\log_2 4^3$.
$\log_2 4^3 = \log_2 (2^2)^3 = \log_2 2^6 = 6 \log_2 2 = 6 \times 1 = 6$
8. Simplify $\log_6 2 + \log_6 3$.
$\log_6 2 + \log_6 3 = \log_6 (2 \times 3) = \log_6 6 = 1$
9. Simplify $\log_5 45 - 2 \log_5 3$.
$\log_5 45 - 2 \log_5 3 = \log_5 45 - \log_5 3^2 = \log_5 45 - \log_5 9$
$= \log_5 (45/9) = \log_5 5 = 1$

10. Simplify $\log_3 63 - 2\log_3 11 + \log_3 242 - \log_3 14$.
The expression $= \log_3 63 - \log_3 121 + \log_3 242 - \log_3 14$

$$= \log_3\left(\frac{63}{121}\right) + \log_3\left(\frac{242}{14}\right) = \log_3\left(\frac{\not{63}^{9}}{\not{121}_{1}} \times \frac{\not{242}^{\not{2}\,1}}{\not{14}_{\not{7}\,1}}\right)$$

$= \log_3 9 = \log_3 3^2 = 2\log_3 3 = 2$

11. Convert to an indicial relationship $\log_4 8 = 3/2$.
The indicial relationship is $8 = 4^{3/2}$. In general, $\log_b N = p$ becomes $N = b^p$.

12. Convert the following into indicial relationships: $\log_5 25 = 2$; $\log_3 81 = 4$; $\log_4 (1/16) = -2$.
They are: $25 = 5^2$; $81 = 3^4$; $1/16 = 4^{-2}$. In general $\log_a P = x$ becomes $P = a^x$.

13. Convert to an indicial relationship $2\log_5 3 + 3\log_5 2 = \log_5 72$.
The equation can be written:

$$\log_5 (3^2 \times 2^3) = \log_5 72$$
$$3^2 \times 2^3 = 72$$

14. Simplify $2\log_a (a^2) + \frac{1}{4}\log_a \sqrt{a} - \frac{3}{2}\log_a (a^3 \times a^{-\frac{1}{4}})$.
It becomes:

$$\begin{aligned} &2 \times (2\log_a a) + \tfrac{1}{4}\log_a a^{\frac{1}{2}} - \tfrac{3}{2}\log_a a^{11/4} \\ &= 4\log_a a + \tfrac{1}{2} \times \tfrac{1}{4}\log_a a - \tfrac{3}{2} \times \tfrac{11}{4}\log_a a \\ &= 4\log_a a + \tfrac{1}{8}\log_a a - \tfrac{33}{8}\log_a a \\ &= (4 + \tfrac{1}{8} - \tfrac{33}{8})\log_a a = 0 \times 1 = 0 \end{aligned}$$

15. Solve the equation $4.3^{x^2-2} = 7.1^{0.4x}$.
Take logs to base 10:

$$(x^2 - 2)\log 4.3 = 0.4x \, . \log 7.1$$

Divide by $\log 4.3$:

$$\begin{aligned} x^2 - 2 &= \frac{0.4 \log 7.1}{\log 4.3} x \\ &= \frac{0.4 \times 0.8512583}{0.6334685} x \\ &= \frac{0.3405033}{0.6334685} x \\ &= 0.5375222x \approx 0.538x \end{aligned}$$

Then $x^2 - 0.538x - 2 = 0$ is the equation to be solved (by the formula):

$$x = \frac{0.538 \pm \sqrt{(0.538)^2 + 8}}{2} = \frac{0.538 \pm \sqrt{0.289444 + 8}}{2}$$

$$= \frac{0.538 \pm \sqrt{8.289444}}{2} = \frac{0.538 \pm 2.8791395}{2}$$

$$= 1.7085698 \text{ or } -1.1705698$$

$$\approx 1.71 \text{ or } -1.17$$

Exercise 2.1

Convert the following indicial equations and statements into their logarithmic counterparts, according to the base indicated:

1. $7^3 = 343$ (a) base 7; (b) base 343; (c) base 10; (d) base 49; (e) base a
2. $a^b = c$ (a) base a; (b) base c; (c) base 10; (d) base a^2; (e) base b
3. $N = ae^{px}$ (a) base e; (b) base 10; (c) base b; (d) base $\left(\frac{N}{a}\right)$
4. $q/q_0 = e^{-t/CR}$ (a) base e; (b) base 10; (c) base b
5. $pv^n = k$ (a) base 10; (b) base e; (c) base v

Convert the following logarithmic statements into indicial statements:

6. $\log_5 125 = 3$
7. $\log_2 512 = 9$
8. $\log_{10} 512 = 9 \log_{10} 2$
9. $\log_4 (1/64) = -3$
10. $\log_{10} 81 = 4 \log_{10} 3$
11. $1 = 4 \log_{81} 3$
12. $\log_b u = c$
13. $\log_b a^2 = p$
14. $c \log_b a = d$
15. $\log_{10} a = c \log_{10} b$
16. $\ln 25 = 2 \ln 5$

Simplify:

17. $10^{\log 3}$
18. $e^{\ln 5}$
19. $e^{\ln a}$
20. $e^{\ln a^2}$

21. e^{21na}
22. $b^{\log_b(1/x)}$
23. $a^{\log_a(N^p)}$
24. $\log_2 32$
25. $\log_2 (1/8)$
26. $2 \log_4 (1/8)$
27. $\log_{\frac{1}{2}} (1/8)$
28. $\log_{0.4} (0.16)$
29. $\log_5 125$
30. $-\frac{5}{6} \log_3 (729)$
31. $\frac{3}{4} \log_5 625$
32. $2 \log_5 5 + 2 \log_5 3 - \log_5 45$
33. $\dfrac{3 \log 2 + \log 4}{2 \log 4 - \log 2}$
34. $\dfrac{2 \log x + \frac{1}{2} \log x^3}{7} - \frac{1}{2} \log x^{\frac{1}{2}}$
35. $\log_4 24 + 2 \log_4 2 + \log_4 100 - \log_4 6$
36. $\log_2 8 + \log_2 5 - 3 \log_2 (1/2) + \log_2 (1/5)$

Solve the following equations:

37. $3^x = 4^2$
38. $4^p = 5^3$
39. $5^{-p} = 6^4$
40. $3^{x+1} = 4^{3-x}$
41. $8^{1+1/x} = 7^{2/x}$
42. $9^{2x+3} = 7^{3x-2}$
43. $8.23^{3x+1} = 6.72^{4x-2}$
44. $32.51^{x+1} = 56.89$
45. $4^x \times 9^{x+2} = 136$
46. $4^{3x-1} = e^{x+2}$
47. $47.09^{2+3x} = 53.65^{3+4x}$
48. $25^{\frac{1}{2}x^2} = 5^{x+2}$
49. $7^x = 5^{x^2-2}$
50. $3.65^{x^2-1} = 6.82^{0.52}$
51. $3^{x^2-8} = 9^{-x}$
52. Taking Newton's law of cooling to be approximately $\theta - \theta_0 = (\theta_1 - \theta_0)e^{-at}$, where θ is the temperature of the body, θ_0 that of the environment, θ_1 the initial temperature of the body and t the time taken for the cooling; a is constant. Determine a if $\theta_0 = 18°$ C, $\theta_1 = 98°$ C, $\theta = 84°$ C and $t = 10$ minutes.
53. Given $V = V_0 e^{pt/25}$, determine p when $t = 6$ and $V = 3V_0$.

54. Given $I = \frac{E}{R}(1 - e^{-Rt/L})$ determine L when $I = 2.4$, $E = 253$, $R = 105$ and $t = 3.2$.
55. A gas expands adiabatically according to the law $pv^{\gamma} = k$. If the initial pressure and volume are 2.78 and 106.5, and the final volume and pressure are 217.0 and 1.26 respectively, determine γ and k.

3

The Solution of Exponential Growth and Decay Problems

The constant e is incommensurable. Its value is approximately 2.7182818285. The following table of values shows how the function e^x changes with x. The values can be checked on a calculator.

x	0	0.5	1.0	1.5
e^x	1	1.6487212	2.71828185	4.4816891

x	10	100	200
e^x	2.20261466	2.6881171×10^{43}	7.2259738×10^{86}

x	-0.5	-2.0	-10.0
e^x	0.6065307	0.1353353	0.0000454

x	-100	-200
e^x	3.720076×10^{-44}	$1.3838965 \times 10^{-87}$

From the above table we notice that as x increases from 0 to $+\infty$, e^x increases from 1 to $+\infty$. As x decreases from 0 to $-\infty$, e^x decreases from 1 to 0; e^x is never negative.

3.1 Definition of e^x

e^x may be defined in many ways. The one adopted here is that the derivative of e^x is itself, i.e.

$$\text{Where } dy/dx = y, \text{ then } y = e^x \qquad 3.1$$

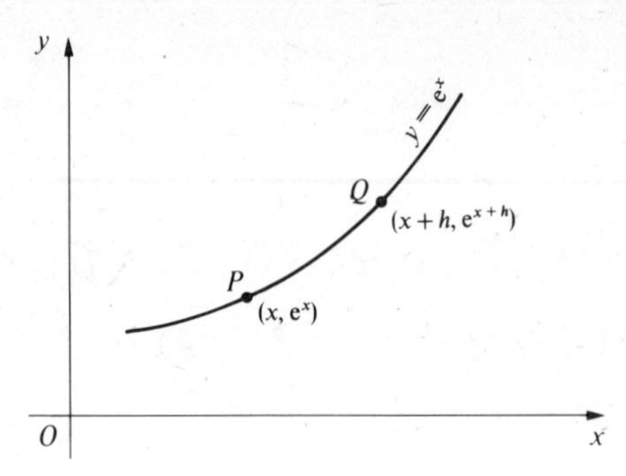

Figure 3.1

A simple calculation serves to show that this is not unreasonable. Fig. 3.1 is a sketch of the curve $y = e^x$; the points P and Q have co-ordinates (x, e^x) and $(x + h, e^{x+h})$ respectively. The gradient of the line PQ is approximately that of the curve at P providing h is small, i.e. Q is close to P.

The gradient of PQ is:

$$\frac{e^{x+h} - e^x}{x + h - x} = \frac{e^x(e^h - 1)}{h}$$

Now give h a small numerical value, say 0.001. The gradient of PQ is:

$$e^x . \frac{(1.0010005 - 1)}{0.001}$$

$$\approx e^x . \frac{0.001}{0.001} = e^x$$

An alternative way by which it can be demonstrated is by assuming that e^x can be expanded in a certain form. In fact:

$$e^x = 1 + x + x^2/2! + x^3/3! + x^4/4! + \ldots$$

Then, by differentiating each term of this we obtain:

$$\frac{d}{dx}(e^x) = 0 + 1 + x + x^2/2! + x^3/3! + x^4/4! + \ldots$$

$$= e^x$$

Formula 3.1 leads to other important results:

Where $y = e^{-x}$, $dy/dx = -e^{-x}$ 3.2

Where $y = e^{ax}$, $dy/dx = ae^{ax}$ 3.3

Where $y = e^{-ax}$, $dy/dx = -ae^{-ax}$ 3.4

3.2 Graphs of e^{ax} and e^{-ax}

Example

Construct a table of values of e^x from 0 to 1 in steps of 0.1 and draw the graph.

x	0	0.1	0.2	0.3	0.4	0.5
e^x	1	1.105	1.221	1.350	1.492	1.649
x	0.6	0.7	0.8	0.9	1.0	
e^x	1.822	2.014	2.226	2.460	2.718	

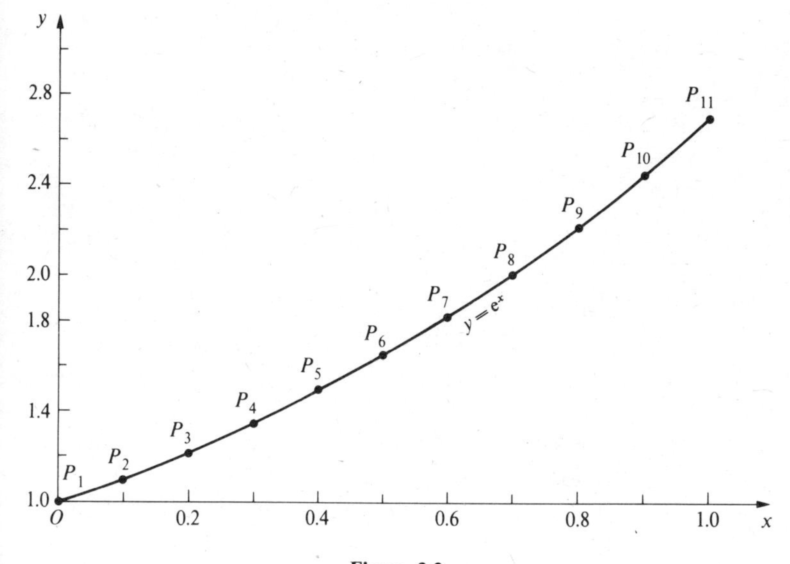

Figure 3.2

From the table Fig. 3.2 is drawn. To estimate the gradient of the curve at a point a tangent may be drawn and its gradient determined graphically. However, it is not easy to draw a line which is an exact tangent to a curve at a given point. Another method of estimating reasonably accurately the gradient of the curve at a particular point is more suitable here. Mark the points plotted for the graph $P_1, P_2, P_3 \ldots P_{11}$. The co-ordinates are (0, 1), (0.1, 1.105), (0.2, 1.1221) . . . (1.0, 2.718). The gradient of the line P_1P_2 is an approximation to the gradient of the curve at P_1. It is an even more accurate approximation to the gradient of the curve at a point between P_1 and P_2 with x co-ordinate = 0.05. Similarly, the gradient of P_2P_3 is approximately the gradient of the curve where $x = 0.15$, and so on. By calculation, the gradient of P_1P_2 is:

$$\frac{y_2 - y_1}{x_2 - x_1} = \frac{1.105 - 1}{0.1} = 1.05$$

The value of e^x, by calculator, when $x = 0.05$ is 1.0512711.
Gradient P_2P_3 is:

$$\frac{1.221 - 1.105}{0.2 - 0.1} = \frac{0.116}{0.1} = 1.16$$

The value of e^x when $x = 0.15$ is 1.1618342.

Gradient P_6P_7 is:

$$\frac{1.822 - 1.649}{0.6 - 0.5} = \frac{0.173}{0.1} = 1.73$$

The value of e^x when $x = 0.55$ is 1.733253.
Now calculate the gradients of P_3P_4, P_4P_5, P_5P_6, P_7P_8, P_9P_{10}, $P_{10}P_{11}$, and check their values against the values of e^x when $x = 0.25, 0.35, 0.45, 0.65, 0.75, 0.85, 0.95$.

Exercise 3.1

Plot the values of e^x for values of x from -10 to 0 in steps of 1 on a graph. Mark the points on the graph P_1, P_2, . . . P_{11} whose x co-ordinates are -10, -9, -8, . . . 0.

Calculate the gradients of P_1P_2, P_2P_3, . . . $P_{10}P_{11}$ and compare their values with the values of e^x when x takes the values -9.5, -8.5, -7.5 . . . -0.5.

Examples

1. Plot on a graph the values of e^{-x} for the values of x from 1 to 10 in steps of 1. Mark the points on the graph P_1, P_2, P_3 . . . P_{10} whose co-ordinates are 1, 2, 3 . . . 10 respectively. Calculate the gradients of P_1P_2, P_2P_3, . . . P_9P_{10} and compare their values with those of $-e^{-x}$ when x takes the values 1.5, 2.5, 3.5 . . . 9.5.

x	1	2	3	4	5
e^{-x}	0.3679	0.1353	0.0498	0.0183	0.0067
x	6	7	8	9	10
e^{-x}	0.0025	0.0009	0.0003355	0.0001234	0.0000454

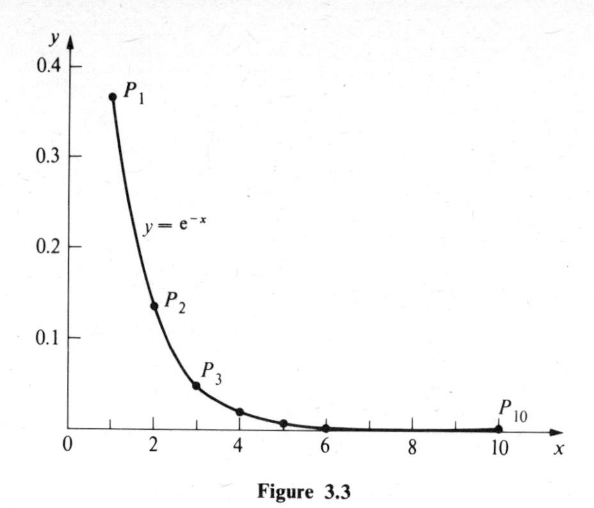

Figure 3.3

In Fig. 3.3 gradient P_1P_2 is:

$$\frac{0.1353 - 0.3679}{2 - 1} = -0.2326$$

Value of $(-e^{-x})$ at $x = 1.5$ is -0.2231.
Gradient P_2P_3 is:

$$\frac{0.0498 - 0.1353}{3 - 2} = -0.0855$$

Value of $(-e^{-x})$ at $x = 2.5$ is -0.082085.
Gradient P_3P_4 is:

$$\frac{0.0183 - 0.0498}{4 - 3} = -0.0315$$

Value of $(-e^{-x})$ at $x = 3.5$ is -0.0301974.
Now calculate the gradients of P_4P_5, P_5P_6 . . . P_9P_{10} and check their values against the values of $-e^{-x}$ at $x = 4.5, 5.5, 6.5, 7.5, 8.5, 9.5$. This time the agreement is not so close as in the previous example, yet the gradients of the lines and the values of $(-e^{-x})$ are in rough agreement. Note that here the intervals between successive values of x are 1, whereas in the previous example the intervals were each 0.1. The successive points were much closer together, so giving a more accurate estimate of the gradient at the mid-interval point.

2. Draw the graph of $y = e^{1.3x}$ for values of x from 2.0 to 3.0 in steps of 0.1. Label the points whose x co-ordinates are 2.0, 2.1, 2.2 . . . 3.0, P_1, P_2,

$P_3 \ldots P_{11}$ respectively. Calculate the gradients of P_1P_2, P_2P_3, $P_3P_4 \ldots P_{10}P_{11}$ respectively, and compare their values with those of $1.3e^{1.3x}$ when x takes the values 2.05, 2.15 . . . 2.95.

x	2.0	2.1	2.2	2.3	2.4	
$e^{1.3x}$	13.464	15.333	17.462	19.886	22.646	
x	2.5	2.6	2.7	2.8	2.9	3.0
$e^{1.3x}$	25.790	29.371	33.448	38.092	43.380	49.402

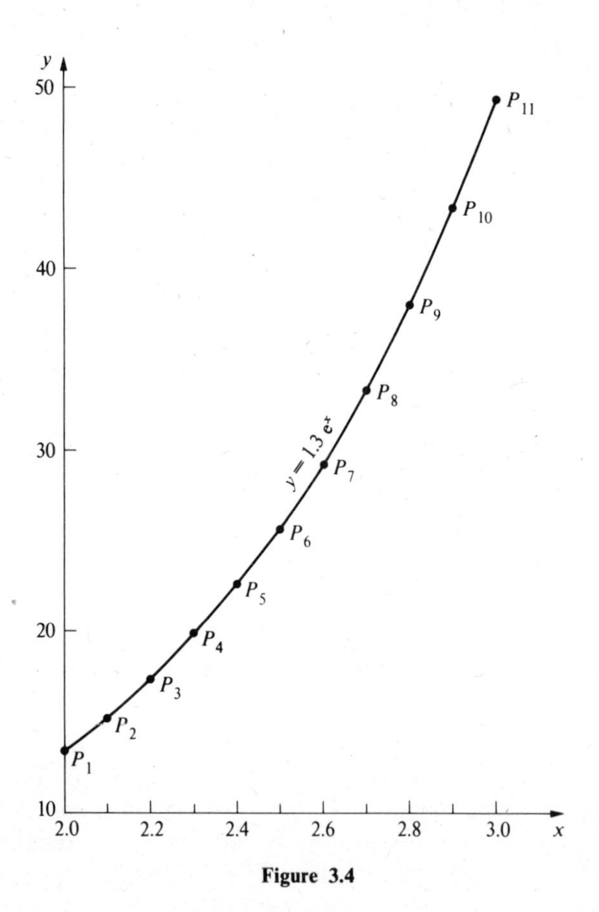

Figure 3.4

In Fig. 3.4 gradient P_1P_2 is:

$$\frac{15.333 - 13.464}{0.1} = 18.69$$

Value of $1.3e^x$ at $x = 2.05 = 18.69$.
Gradient P_2P_3 is:

$$\frac{17.462 - 15.333}{0.1} = 21.29$$

Value of $1.3e^{1.3 \times 2.15} = 21.271$.
Gradient P_3P_4 is:

$$\frac{19.886 - 17.462}{0.1} = 24.24$$

Value of $1.3e^{1.3 \times 2.25} = 24.224$.
Now complete the rest of the calculations. Note that there is close agreement between the gradients of the chords P_1P_2, P_2P_3, etc. and the values of $1.3e^{1.3x}$ at the mid-interval values of x.

Exercise 3.2

Draw the graphs of the following functions for the range of values of x given. In similar fashion to the examples already studied, check that the estimated gradient of the curve agrees reasonably closely with the appropriate values given by formula 3.3 or 3.4, whichever is relevant.

1. $y = e^{0.75x}$ from $x = 1.0$ to 2.0 in steps of 0.1
2. $y = e^{-0.25x}$ from $x = -2.0$ to -1.0 in steps of 0.1
3. $y = e^{1.8x}$ from $x = 3.0$ to 4.0 in steps of 0.1
4. $y = e^{-2.1x}$ from $x = -3.0$ to -2.0 in steps of 0.1
5. $y = e^{3.2x}$ from $x = 2.0$ to 2.1 in steps of 0.01
6. $y = e^{-5.6x}$ from $x = -1.1$ to -1.0 in steps of 0.01

3.3 Growth and decay problems

In certain problems in biology the rate of increase of a population is proportional to the size of the population. In such cases we are involved in the growth of the population. In certain aspects of radioactivity and in the discharge of a capacitor we are concerned with the issue of the decay of a mass or of a charge. Such problems are very much concerned with the exponential function, that is, some form of e^{ax}. The following examples illustrate these ideas.

Examples

1. The mass of a radioactive substance, m kilograms, at a given moment in time, t seconds, is related with the mass m_0 kilograms at time $t = 0$ by means of the following equation:

$$m = m_0 e^{-pt} \qquad 3.5$$

The constant, p, is called the radioactive decay constant. A radium isotope has a decay constant $1.359 \times 10^{-11} s^{-1}$. Plot the graph of m for the interval 0 to 10^{11} seconds in intervals of 10^{10} seconds. From the graph determine the half life of the substance, i.e. the time for one half of the original mass to have been radiated. Take m_0 to be 1 kg.

t	0	10^{10}	2×10^{10}	3×10^{10}	4×10^{10}
pt	0	1.359×10^{-1}	2.718×10^{-1}	4.077×10^{-1}	5.436×10^{-1}
e^{-pt}	1	0.873	0.762	0.665	0.581
m(kg)	1	0.873	0.762	0.665	0.581

t	5×10^{10}	6×10^{10}	7×10^{10}	8×10^{10}
pt	6.79×10^{-1}	8.15×10^{-1}	9.513×10^{-1}	1.087
e^{-pt}	0.507	0.442	0.386	0.337
m(kg)	0.507	0.442	0.386	0.337

t	9×10^{10}	$10 \times 10^{10} (= 10^{11})$
pt	1.223	1.359
e^{-pt}	0.294	0.257
m(kg)	0.294	0.257

From the graph, Fig. 3.5, the half life is approximately 5.1×10^{10}. The approximate rate of change of mass at 3.5×10^{10} seconds is calculated to be:

$$\frac{0.581 - 0.665}{4 \times 10^{10} - 3 \times 10^{10}} = -\frac{0.084}{10^{10}} = 8.4 \times 10^{-12} \text{ kg/s}$$

By formula *3.4* it should be:

$$-pe^{-p \times 3.5 \times 10^{10}}$$

$$= -1.359 \times 10^{-11} \times (e^{-1.359 \times 3.5 \times 10^{-1}})$$

$$= -0.8445926 \times 10^{-11} = -8.4 \times 10^{-12} \text{ kg/s}$$

To check on the above half life calculations, substituting in formula *3.5*:

$$\frac{m_0}{2} = m_0 e^{-pT_{\frac{1}{2}}}, \text{ where } T_{\frac{1}{2}} \text{ is the half life.}$$

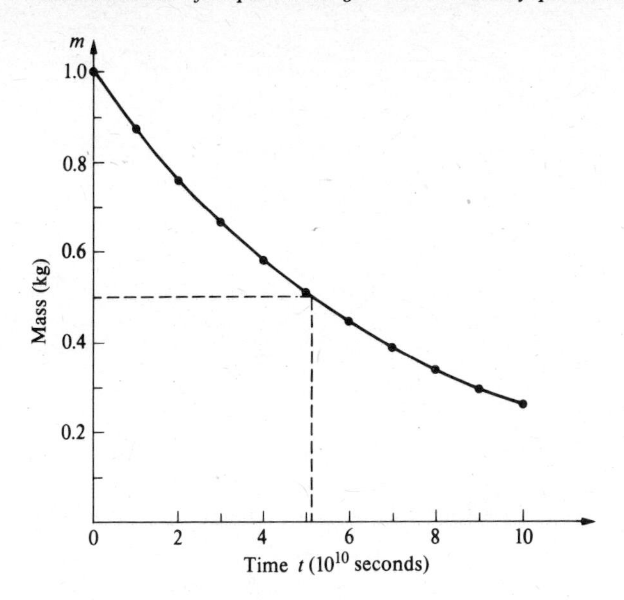

Figure 3.5

i.e. $\frac{1}{2} = e^{-pT_{\frac{1}{2}}}$

Take natural logarithms $\ln(0.5) = -pT_{\frac{1}{2}}$

Then
$$T_{\frac{1}{2}} = -\frac{1}{1.359 \times 10^{-11}} \times \ln(0.5)$$
$$= -\frac{1}{1.359 \times 10^{-11}}(-0.6931)$$
$$= \frac{0.510}{10^{-11}} = 5.1 \times 10^{10}$$

2. The formula $T = T_0 e^{\mu\theta}$ is used to relate the tensions in a belt drive. T_0 is the tension in the slack side of the belt and T is that at any point of the belt in contact with the pulley; μ is the coefficient of friction between the belt and the pulley and θ is the angle of lap around the pulley, in radians (Fig. 3.6).

   Draw a graph for T as θ varies from 0 to 2.8 radians in steps of 0.4 radians. Take μ to be approximately the value 0.4 and T_0 to be 50 N. Determine the approximate value of T_1 from the graph, assuming the total lap to be 2.65 radians. Check this value by a direct calculation from the formula. Determine the approximate rate of increase of T when

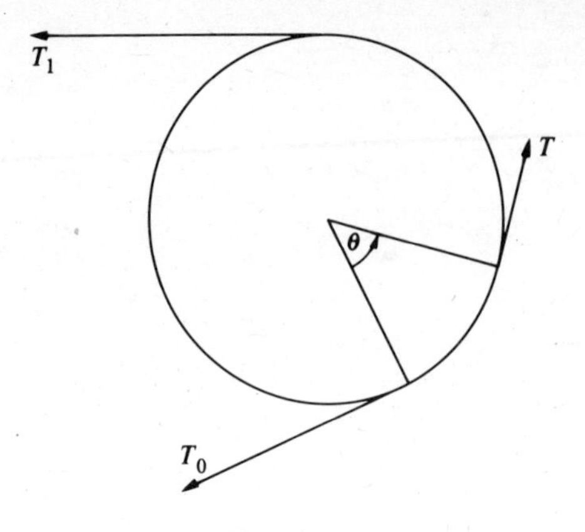

Figure 3.6

$\theta = 1.4$ radians. Check this value by a calculation based on formula *3.3*.

θ	0	0.4	0.8	1.2	1.6	2.0
$\mu\theta$	0	0.16	0.32	0.48	0.64	0.8
$e^{\mu\theta}$	1	1.174	1.377	1.616	1.896	2.226
T	50	58.7	68.8	80.8	94.8	111.3

θ	2.4	2.8
$\mu\theta$	0.96	1.12
$e^{\mu\theta}$	2.612	3.065
T	130.6	153.2

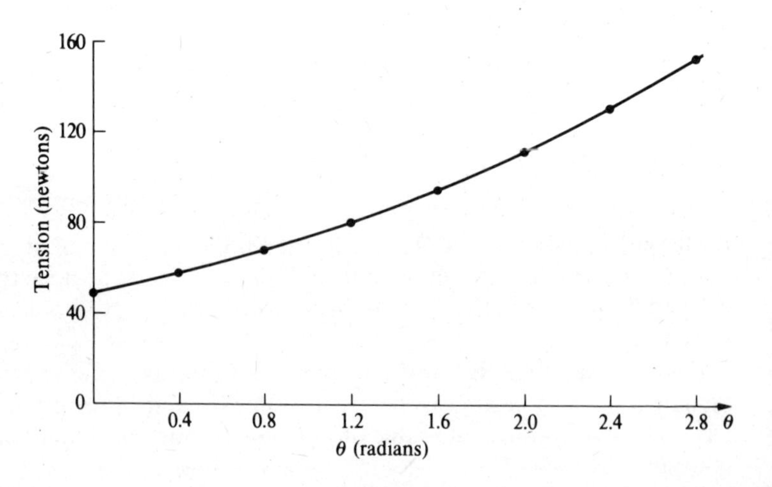

Figure 3.7

From the graph, Fig. 3.7, $T_1 \approx 144$ N. By calculation:

$$T_1 = 50e^{0.4 \times 2.65} = 50 \times 2.886371$$
$$= 144.31855 \approx 144 \text{ N}$$

The approximate rate of increase when $\theta = 1.4$ radians is:

$$\frac{94.8 - 80.8}{1.6 - 1.2} = 35 \text{ N/rad}$$

From formula *3.3* the rate of increase is:

$$T_0 \mu e^{\mu\theta} = 50 \times 0.4e^{0.4 \times 1.4} = 35.01345 \approx 35 \text{ N/rad}$$

3.

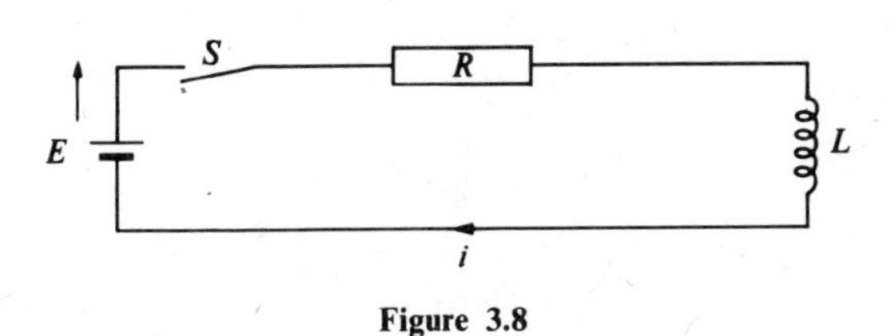

Figure 3.8

In a circuit such as the one illustrated in Fig. 3.8 where a resistor, R ohms, and an inductance, L henry, are connected in series to an electromotive force, E volts, by means of a switch, S, the current, i, at the moment t seconds after S is closed is given by:

$$i = \frac{E}{R}(1 - e^{-Rt/L})$$

Draw the graph for i where $E = 2.2$ V, $R = 10.5\Omega$, and $L = 0.25$ H, for the values of t from 0 to 0.12 seconds in steps of 0.02 seconds.

From the graph determine the value of i when $t = 0.07$ s and the value of t when $i = 0.15$ A. Determine the approximate rate of change of i when $t = 0.05$ and 0.03 seconds. Check these answers by a direct calculation.

t	0	0.02	0.04	0.06	0.08	0.10	0.12
Rt	0	0.21	0.42	0.63	0.84	1.05	1.26
Rt/L	0	0.84	1.68	2.52	3.36	4.20	5.04
$e^{-Rt/L}$	1	0.432	0.186	0.080	0.035	0.015	0.006
$1 - e^{-Rt/L}$	0	0.568	0.814	0.920	0.965	0.985	0.994
i	0	0.12	0.17	0.19	0.20	0.21	0.21

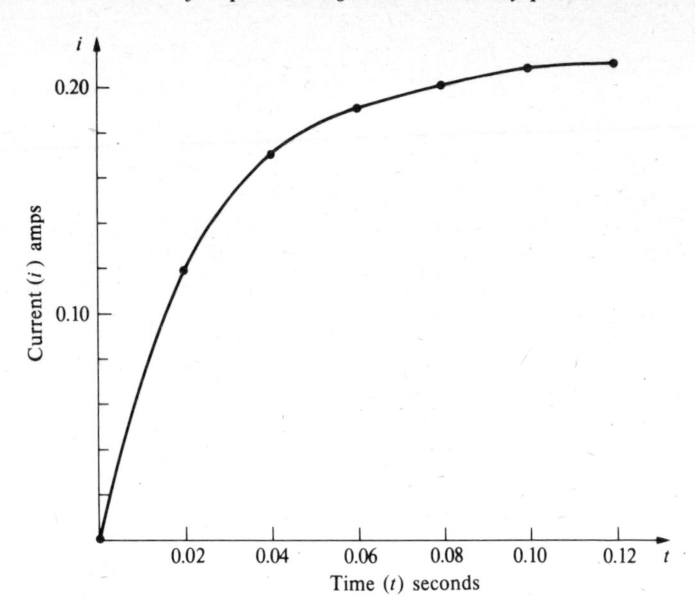

Figure 3.9

From the graph, Fig. 3.9, when $t = 0.07$, $i \approx 0.20$.

By calculation, $i = \dfrac{2.2}{10.5}(1 - e^{-2.94}) = 0.1984472 \approx 0.20$.

Rate of change of $i(t = 0.05)$: $= \dfrac{0.19 - 0.17}{0.06 - 0.04} = \dfrac{0.02}{0.02} = 1.$

Rate of change of $i(t = 0.07)$: $= \dfrac{0.20 - 0.19}{0.08 - 0.06} = \dfrac{0.01}{0.02} = 0.5.$

By calculation, the rate of change of $i = \dfrac{E}{R} \cdot \dfrac{R}{L} e^{-Rt/L}$.

When $t = 0.05$, $= \dfrac{2.2}{0.25} e^{-10.5 \times 0.05/0.25} = \dfrac{2.2}{0.25} \times 0.1224564 \approx 1.078.$

When $t = 0.07$, $= \dfrac{2.2}{0.25} e^{-10.5 \times 0.07/0.25} = \dfrac{2.2}{0.25} \times 0.0528657 \approx 0.465.$

4. A capacitor is discharged slowly through a resistor. If the capacitance of the capacitor is C farads and the resistance of the resistor is R ohms, the quantity of charge, Q coulombs, at time t seconds, on the plate of the capacitor is given by $Q = Q_0 e^{-t/CR}$, where Q_0 is the initial charge on the plate.

   Sketch the graph of Q against t. Determine the value of Q when $t = 1.5 \times 10^{-3}$ s, given $Q_0 = 2.5 \times 10^{-4}$ C, $R = 2 \times 10^3\ \Omega$,

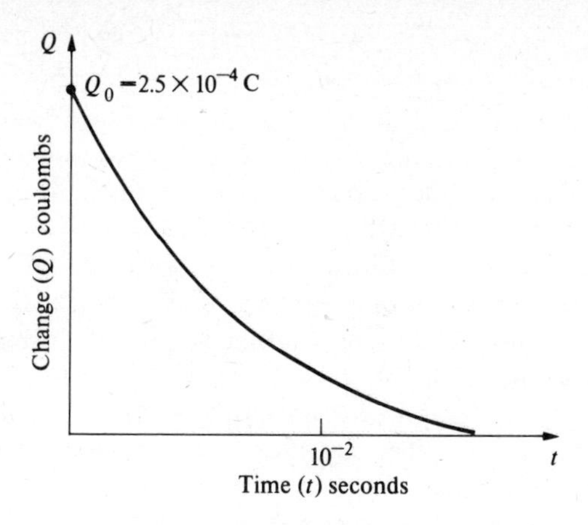

Figure 3.10

$C = 3 \times 10^{-6}$ F, and the value of t when $Q = 1.2 \times 10^{-4}$ C. Also determine the discharging current (i.e. dQ/dt) when $t = 0.75 \times 10^{-3}$ s.

See Fig. 3.10. When $t = 1.5 \times 10^{-3}$:

$$Q = 2.5 \times 10^{-4} e^{-1.5/6} = 2.5 \times 10^{-4} \times e^{-0.25}$$

$$= 2.5 \times 10^{-4} \times 0.7788008 = 1.947002 \times 10^{-4} \approx 1.95 \times 10^{-4}$$

From $Q/Q_0 = e^{-t/CR}$ we obtain:

$$-t/CR = \ln(Q/Q_0)$$
$$t = -CR \ln(Q/Q_0)$$

When $Q = 1.2 \times 10^{-4}$:

$$\begin{aligned} t &= -3 \times 10^{-6} \times 2 \times 10^{3} \times \ln(1.2 \times 10^{-4}/2.5 \times 10^{-4}) \\ &= -6 \times 10^{-3} \times \ln(1.2/2.5) = -6 \times 10^{-3} \times (-0.7339692) \\ &= 4.4038151 \times 10^{-3} \approx 4.4 \times 10^{-3}\text{s} \end{aligned}$$

$dQ/dt = Q_0 \times \left(-\dfrac{1}{CR}\right)e^{-t/CR} = -\dfrac{Q_0}{CR}e^{-t/CR}$. When $t = 0.75 \times 10^{-3}$:

$$\frac{dQ}{dt} - \frac{2.5 \times 10^{-4}}{3 \times 10^{-6} \times 2 \times 10^{3}} \times e^{-0.75/6}$$

$$= -\frac{2.5}{6} \times 10^{-1} \times e^{-0.75/6} = -\frac{2.5}{6} \times 10^{-1} \times 0.8824969$$

$$= -0.367707 \times 10^{-1} \approx -3.68 \times 10^{-2}\text{ A}$$

Exercise 3.3

1. From the formula for radioactive decay $M = M_0 e^{-at}$, where $M_0 = 1.5$, $a = 8.47 \times 10^{-10}\,s^{-1}$, plot a graph for M for the interval 0 to 10^{10} s in steps of 10^9 s. From the graph estimate (a) the half life of the substance, (b) the rate of the decay when $t = 0.55 \times 10^9$ s, (c) the value of M when $t = 0.25 \times 10^9$ s. Check all three results by calculation.
2. During a certain period of a plant's growth its height H is given by the formula $H = H_0 e^{pt}$, where $H_0 = 10$ cm and $p = 0.07$, and t is measured in days. Plot the graph for H for the interval $t = 5$ to $t = 15$ in steps of 1 day. From the graph estimate (a) when $H = 2H_0$, (b) the rate of growth in centimetres per day when $t = 10$, (c) the value of H when $t = 14$. Check all three results by calculation.
3. From the formula $Q = Q_0 e^{-t/CR}$, given $Q_0 = 7.5 \times 10^{-4}$ C, $R = 5 \times 10^2\,\Omega$, $C = 4 \times 10^{-5}$ F, draw a graph of Q for the interval $t = 0.02$ to 0.04 seconds in steps of 0.002 seconds. From the graph estimate (a) the value of Q when $t = 0.027$ seconds, (b) the discharging current (dQ/dt) when $t = 0.031$ seconds, (c) the value of t when $Q = 3.2 \times 10^{-4}$ C. Check all results by calculation.
4. From the formula $I = I_0 e^{-Rt/L}$, given $I_0 = 2.65$ A, $R = 12.5\,\Omega$ and $L = 0.35$ H, determine (a) the value of I when $t = 0.07$ seconds, (b) the rate of change of I when $t = 0.05$ seconds, (c) the value of t when $I = 1.26$ *A*.
5. From the formula $T = T_0 e^{\mu\theta}$, given $\mu = 0.45$, $T_0 = 195$ N, determine (a) the value of T when $\theta = 2.54$ rad, (b) the rate of increase of T with θ when $\theta = 1.83$ rad, (c) the value of θ when $T = 285$ N.
6. For Newton's law of cooling $\theta = \theta_0 e^{-kt}$. Given $\theta_0 = 235°$ C, $k = 0.0072$, determine (a) θ when $t = 25$, (b) the rate of cooling when $t = 32$, (c) the value of t when $\theta = 149°$ C.
7. Assuming that the relationship between the length of a metal rod, L, and its temperature, θ, is given by $L = L_0 e^{k\theta}$, where $L_0 = 105$ cm and $k = 2.65 \times 10^{-7}$, determine (a) the value of L when $\theta = 85°$ C, (b) the rate of increase of L with θ when $\theta = 70°$ C, (c) the value of θ to produce $L = 106.7$ cm.
8. Assuming the formula for the number of bacteria, n, in a given culture is given by $n = n_0 e^{kt}$, where t is measured in minutes, and when $n_0 = 3.2 \times 10^6$ and $k = 0.015$, determine (a) the value of n when $t = 75$ minutes, (b) the rate of increase per minute in the population when $t = 45$ minutes, (c) the value of t when $n = 4.6 \times 10^9$.

4

Simple Curves. Non-linear Physical Laws

Before the graph of a non-linear curve can be drawn a suitable table of values must be constructed.

4.1 The parabola

Examples

1. Plot the graph of $y = \frac{3}{4}x^2$ between $x = -4$ and $x = 4$.

Step 1.	x	-4	-3	-2					
Step 2.	x^2	16	9	4					
Step 3.	$\frac{3}{4}x^2$	12	$6\frac{3}{4}$	3					

x	-1	0	1	2	3	4
x^2	1	0	1	4	9	16
$\frac{3}{4}x^2$	$\frac{3}{4}$	0	$\frac{3}{4}$	3	$6\frac{3}{4}$	12

Step 4. Choose a suitable scale for x.
Step 5. Note the values of y are symmetrical about $x = 0$.
Step 6. Note that y is never negative.
Step 7. Choose a suitable scale for y. (See Fig. 4.1.)
Step 8. When $x = 1.5$, from the graph $y \approx 1.7$. By calculator, when $x = 1.5$, $y = 0.75 \times 1.5^2 = 1.6875$.
Step 9. When $y = 4.8$, from the graph $x \approx \pm 2.5$. By calculator $x = \pm\sqrt{\frac{4}{3} \times 4.8} = \pm 2.5298221$.

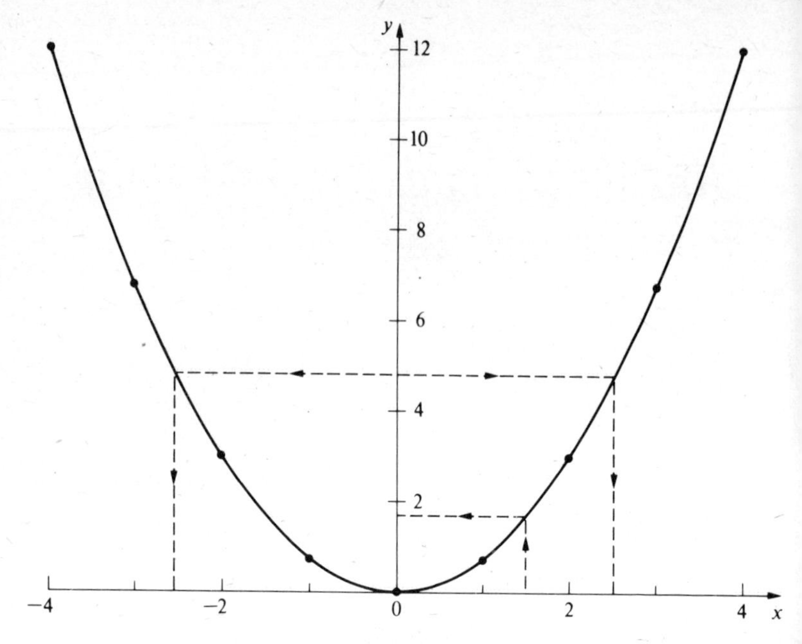

Figure 4.1

2. Plot the graph of $y = -3(x+2)(x-1)$ between $x = -5$ and $x = 3$.

Step 1.	x	-5	-4	-3				
Step 2.	$(x-1)$	-6	-5	-4				
Step 3.	$(x+2)$	-3	-2	-1				
Step 4.	$(x+2)(x-1)$	18	10	4				
Step 5.	$-3(x+2)(x-1)$	-54	-30	-12				
	x	-2	-1	0	1	2	3	
	$(x-1)$	-3	-2	-1	0	1	2	
	$(x+2)$	0	1	2	3	4	5	
	$(x+2)(x-1)$	0	-2	-2	0	4	10	
	$-3(x+2)(x-1)$	0	6	6	0	-12	-30	

Step 6. Choose a suitable scale for x.
Step 7. Note the values of y are symmetrical about $x = -0.5$.
Step 8. Choose a suitable scale for y. (See Fig. 4.2.)
Step 9. When $x = -3.2$, from the graph, $y \approx -15$. By calculator, $y = -3 \times -1.2 \times -4.2 = -15.12$.
Step 10. When $x = -0.5$, by calculator, $y = -3 \times 1.5 \times -1.5 = 6.75$.
Step 11. When $y = -20$, from the graph, $x \approx 2.5$ or -3.5.

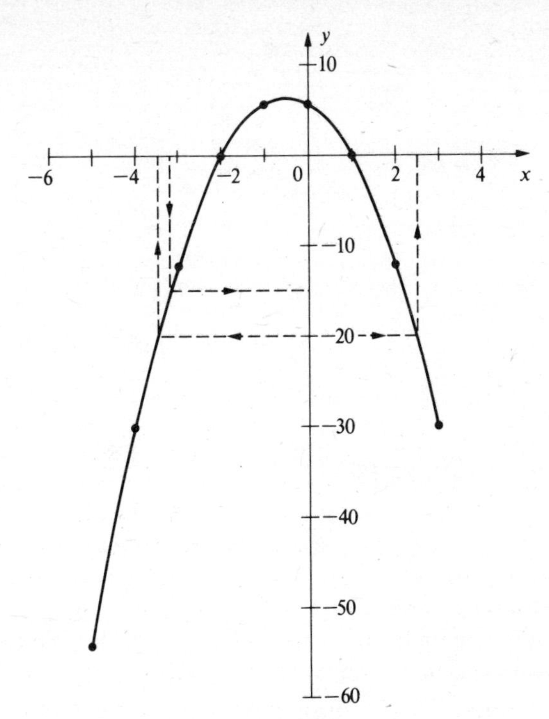

Figure 4.2

3. Plot the graph of $y = 2x^2 - 5x + 11$ between $x = -3$ and $x = 5$.

Step 1.	x	-3	-2	-1
Step 2.	x^2	9	4	1
Step 3.	$2x^2$	18	8	2
Step 4.	$-5x$	15	10	5
Step 5.	11	11	11	11
Step 6.	y	44	29	18

x	0	1	2	3	4	5
x^2	0	1	4	9	16	25
$2x^2$	0	2	8	18	32	50
$-5x$	0	-5	-10	-15	-20	-25
11	11	11	11	11	11	11
y	11	8	9	14	23	36

Step 7. Choose a suitable scale for x.
Step 8. Choose a suitable scale for y. (See Fig. 4.3.)
Step 9. From the above table it does not appear that the curve is symmetrical about any value of x.

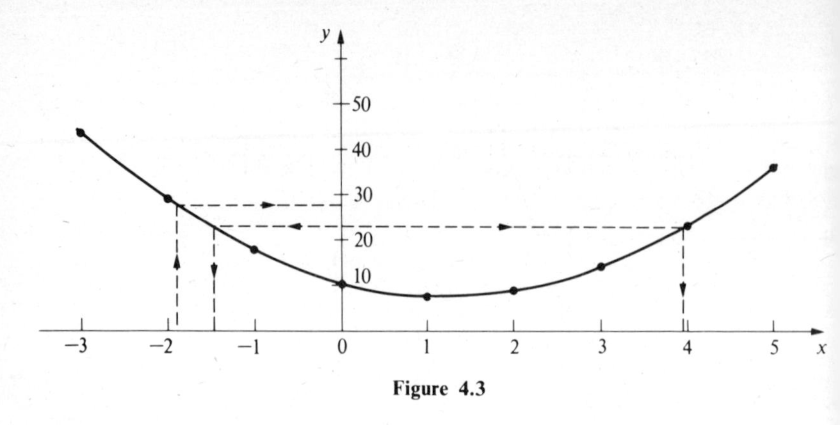

Figure 4.3

Step 10. When $x = -1.8$, from the graph, $y \approx 26$. By calculator, when $x = -1.8$, $y = 26.48$.
Step 11. When $y = 23$, from the graph, $x \approx 3.9$ or -1.4.
Step 12. From the graph the curve is symmetrical about $x = 1.25$.

Each of the above curves is a parabola. The curves in Figs. 4.1 and 4.3 are parabolas with minimum points. The curve in Fig. 4.2 is a parabola with a maximum point. The minimum or maximum point of such a parabola occurs at that value of x about which the curve is symmetrical. The point is often called the vertex.

The basic parabola of this type has an equation $y = x^2$

It has a minimum point (vertex) at (0, 0). y is never negative.

Curves whose equations are $y = ax^2 + bx + c$ are all parabolas. The graphs can all be derived from that of the basic parabola by suitable transformations. Fig. 4.4 is a sketch of $y = x^2$.

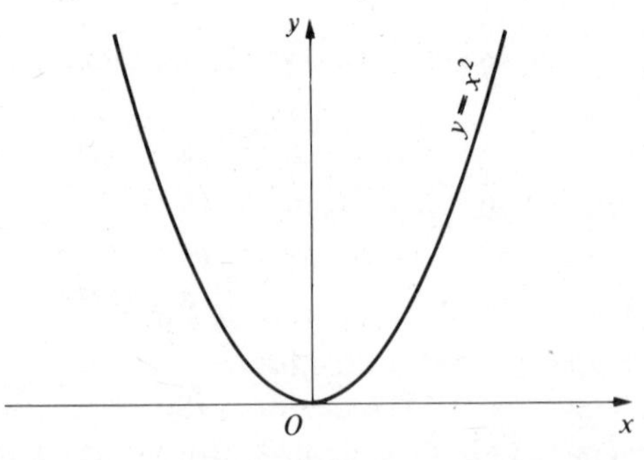

Figure 4.4

he parabola $y = Ax^2$

'or any value of x the value of y in the parabola $y = Ax^2$ equals A times the orresponding value of y in the parabola $y = x^2$. Fig. 4.5 represents the elation between the graphs of $y = x^2$ and $y = Ax^2$ when A is positive and > 1. In Fig. 4.5 an ordinate of $y = Ax^2$ is always A times the correspond-ng ordinate of $y = x^2$.

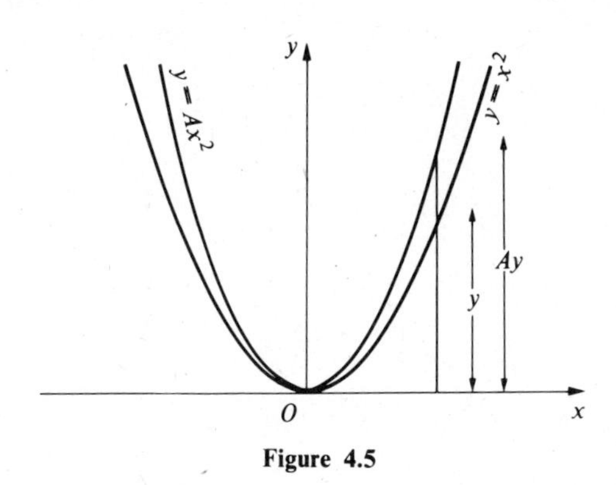

Figure 4.5

Fig. 4.6 represents the relation between the two curves when A is positive nd $A < 1$.

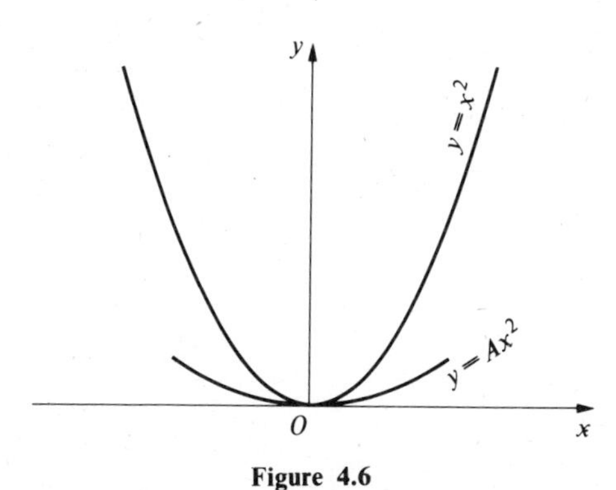

Figure 4.6

Fig. 4.7 represents the relation between the two curves when A is negative nd $|A| > 1$.

Fig. 4.8 represents the relation between the two curves when A is negative nd $|A| < 1$.

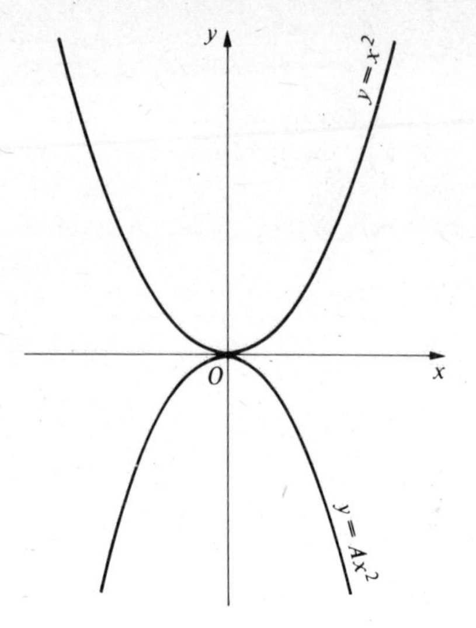

Figure 4.7

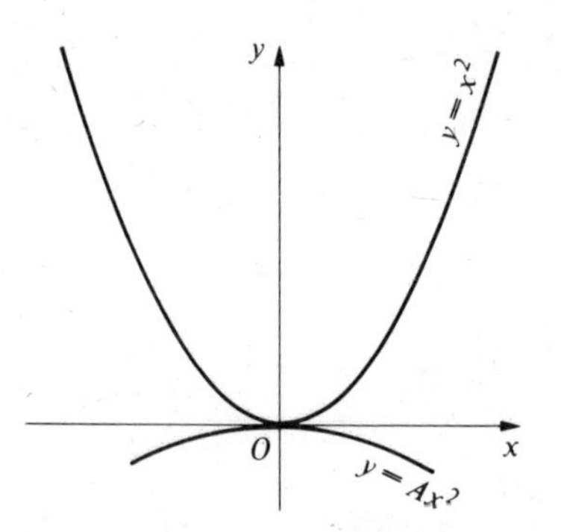

Figure 4.8

The parabola $y = (x-h)^2$

The curve has its minimum point (vertex) when $x = h$, i.e. at the point (h, (

The curve has symmetry about $x = h$. Fig. 4.9 represents the relatio between the curves $y = (x-h)^2$ and $y = x^2$. The curve $y = (x-h)^2$ merely $y = x^2$ moved h units parallel to Ox.

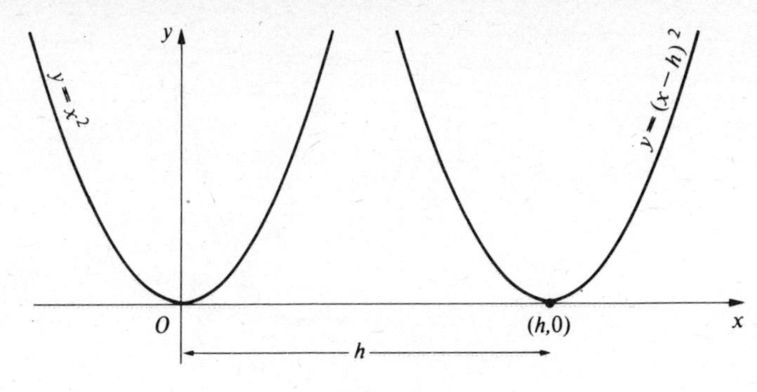

Figure 4.9

he parabola $y = x^2 + k$

he above curve has a rearranged equation $y - k = x^2$. It is symmetrical bout $x = 0$. Its minimum point is $(0, k)$.

Fig. 4.10 represents the relation between the curves $y = x^2$ and $= x^2 + k$. The second parabola is merely the first parabola moved k units arallel to Oy.

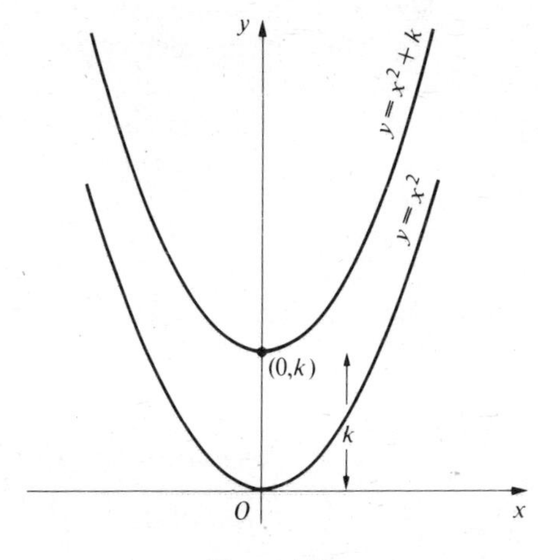

Figure 4.10

ne parabola $y = A(x-h)^2 + k$

he above curve has a rearranged equation $y - k = A(x-h)^2$. This curve ay be derived from the curve $y = x^2$ by three transformations (see ig. 4.11).

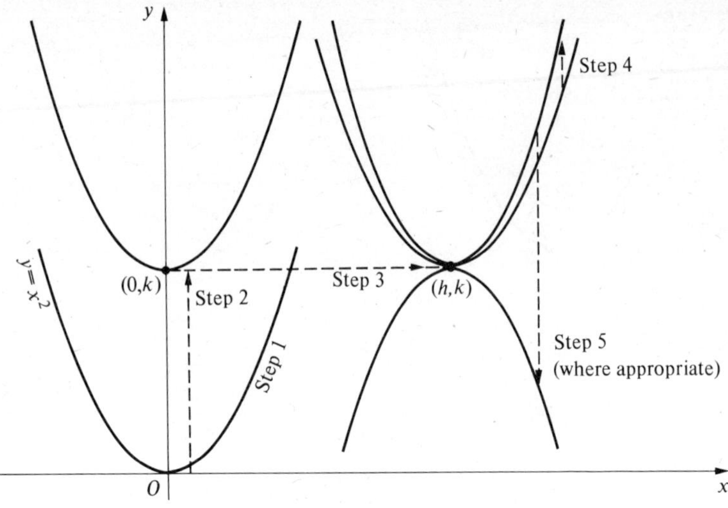

Figure 4.11

Step 1. Draw $y = x^2$.
Step 2. Move the whole curve k units along Oy.
Step 3. Now move the curve h units parallel to Ox.
Step 4. Stretch or compress the curve in the direction of the y ax according to whether $A > 1$ or $A < 1$ and is positive.
Step 5. When A is also negative reflect the curve in the line through i vertex, which is parallel to Ox.

The parabola $y = ax^2 + bx + c$

To determine the relation between the parabola $y = ax^2 + bx + c$ ar the basic parabola $y = x^2$: $y = ax^2 + bx + c$ is related to the parabo $y = A(x - h)^2 + k$:

$$y = a\left(x^2 + \frac{b}{a}x\right) + c$$

$$= a\left(x^2 + \frac{b}{a}x + \frac{b^2}{4a^2}\right) + c - \frac{b^2}{4a}$$

$$= a\left(x + \frac{b}{2a}\right)^2 + \left(\frac{4ac - b^2}{4a}\right)$$

By relating equation (i) to the form $y = A(x-h)^2 + k$ we get:

$$k = \frac{4ac - b^2}{4a}$$

$$h = -\frac{b}{2a}$$

$$A = a$$

Examples

1. Sketch the graph of $y = 3x^2 - 6x + 3$.
   Step 1. Rearrange the equation: $y = 3(x^2 - 2x + 1) = 3(x-1)^2$.
   Step 2. This equation is of the form $y = A(x-h)^2$.
   Fig. 4.12 represents the steps taken to arrive at a sketch of the curve.

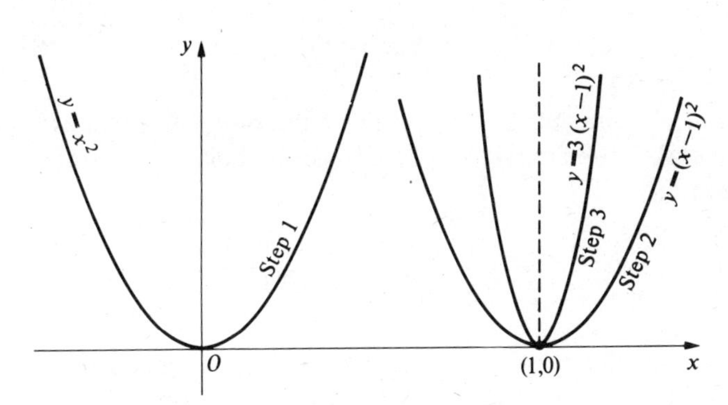

Figure 4.12

2. Sketch the graph of $y = 8 + 5x - 2x^2$.
   Step 1. Rearrange the equation:

$$\begin{aligned} y &= -2\left(x^2 - \tfrac{5}{2}x\right) + 8 \\ &= -2\left(x^2 - \tfrac{5}{2}x + \tfrac{25}{16}\right) + 8 + \tfrac{25}{8} \\ &= -2\left(x - \tfrac{5}{4}\right)^2 + \tfrac{89}{8} \end{aligned}$$

Step 2. $k = 89/8 = 11\tfrac{1}{8}$

Step 3. $h = 5/4$

Step 4. $A = -2$

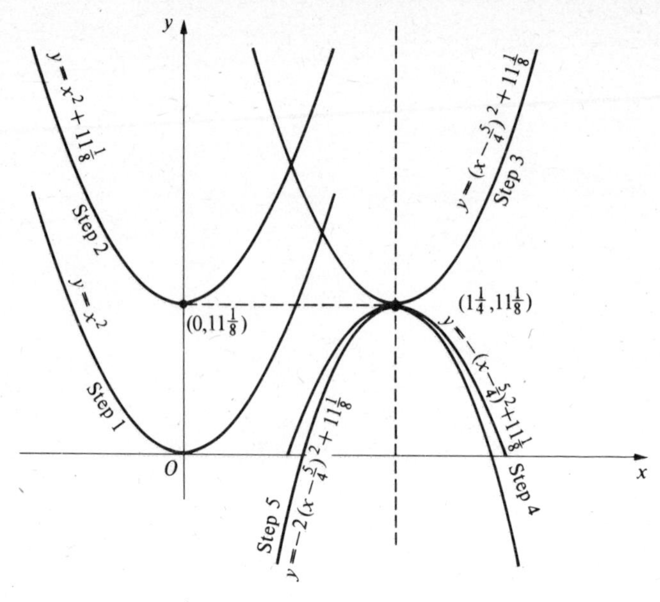

Figure 4.13

Fig. 4.13 represents a sketch of the graph of $y = 8 + 5x - 2x^2$ and the steps taken from the sketch of the basic parabola $y = x^2$ to arrive at the graph.

Exercise 4.1

Construct suitable tables of values and plot the graphs of the following equations between the given values of x. Use the graphs to determine the values of y or x indicated. Wherever possible check those values by calculator.

1. $y = \frac{1}{2}x^2$; between $x = -4$ and $x = 4$. Determine y when $x = 2\frac{1}{2}$ and x when $y = 5\frac{1}{2}$.
2. $y = 4(x-1)(x-2)$; between $x = -5$ and $x = 3$. Determine y when $x = -1.6$ and x when $y = 3.2$.
3. $y = -\frac{1}{5}(x-1)(x-2)$; between $x = -3$ and $x = 4$. Determine y when $x = 1.7$ and x when $y = -1.1$.
4. $y = -\frac{1}{2}x^2 + 2$; between $x = -3$ and $x = 4$. Determine y when $x = -1.3$ and x when $y = 1.5$.
5. $y = x^2 - 5x + 6$; between $x = -1$ and $x = 6$. Determine y when $x = 2.5$ and x when $y = 1.5$.

6. $y = x^2 + 4x + 1$; between $x = -4$ and $x = 2$. Determine where the curve crosses Oy and Ox.
7. $y = 2x^2 - 3x + 2$; between $x = -3$ and $x = 4$. Determine the axis of symmetry of the curve and the co-ordinates of the vertex.
8. $2y + 3 = 6x - 3x^2$; between $x = -1$ and $x = 6$. Determine the axis of symmetry for the curve and the co-ordinates of the vertex.

Sketch the following curves by first rearranging their equations, where necessary, into a suitable form. Show the steps taken to arrive at the final sketch.

9. $y = 3x^2$
10. $y = -\frac{4}{5}x^2$
11. $y = 2(x-1)^2$
12. $y = -\frac{1}{3}(x+2)^2$
13. $y = 2(4-x)^2$
14. $y = x^2 + 3$
15. $y = x^2 - 6$
16. $y = -\frac{3}{4}x^2 + \frac{5}{8}$
17. $y = (x-2)^2 + 5$
18. $3y = -2(x+1)^2 - 4$
19. $y = \frac{4}{3}(x - \frac{2}{5})^2 + \frac{11}{6}$
20. $y = p(x-q)^2 + r$
21. $y = x^2 - 2x + 3$
22. $y = 4x^2 + 12x - 3$
23. $y = x^2 - x + 1$
24. $x = 2(y-1)^2 + 3$

4.2 Further curves

Examples

1. Draw the graph of $y = 2/x$ between $x = -3$ and $x = 3$.

Step 1.	x	-3	-2	-1	-0.5	-0.25	0.25	0.5	1	2	3
Step 2.	$1/x$	$-\frac{1}{3}$	$-\frac{1}{2}$	-1	-2	-4	4	2	1	$\frac{1}{2}$	$\frac{1}{3}$
Step 3.	$2/x$	$-2/3$	-1	-2	-4	-8	8	4	2	1	2/3

Step 4. Choose a suitable scale for x.
Step 5. Note the values of y are asymmetrical about $x = 0$, i.e. they are numerically equal in value but opposite in sign.

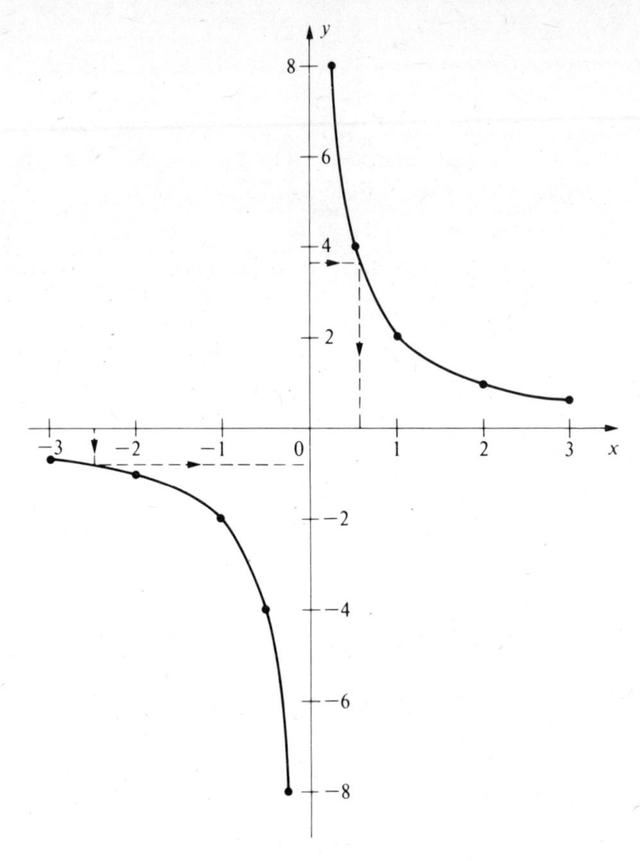

Figure 4.14

Step 6. Choose a suitable scale for y. (See Fig. 4.14.)
Step 7. When $x = -2.5$, from the graph, $y \approx -0.8$. By calculator, $y = -0.8$.
Step 8. When $y = 3.8$, from the graph, $x \approx 0.5$. By calculator, $x = 0.5263158$.

2. Draw the graph of $y = -\dfrac{3}{(2x-5)}$ between $x = -1$ and $x = 6$.

Step 1.	x	-1	0	1	1.5	2
Step 2.	$2x$	-2	0	2	3	4
Step 3.	$(2x-5)$	-7	-5	-3	-2	-1
Step 4.	$\dfrac{1}{2x-5}$	$-\frac{1}{7}$	$-\frac{1}{5}$	$-\frac{1}{3}$	$-\frac{1}{2}$	-1
Step 5.	$\dfrac{-3}{2x-5}$	$\frac{3}{7}$	$\frac{3}{5}$	1	$\frac{3}{2}$	3

x	2.25	2.75	3	3.5	4	5	6
$2x$	4.5	5.5	6	7	8	10	12
$(2x-5)$	-0.5	0.5	1	2	3	5	7
$\frac{1}{2x-5}$	-2	2	1	1/2	1/3	1/5	1/7
$\frac{-3}{2x-5}$	6	-6	-3	$-\frac{3}{2}$	-1	$-\frac{3}{5}$	$-\frac{3}{7}$

Step 6. Choose a suitable scale for x.
Step 7. Note the values of y are asymmetrical about $x = 2.5$.
Step 8. Choose a suitable scale for y. (See Fig. 4.15.)

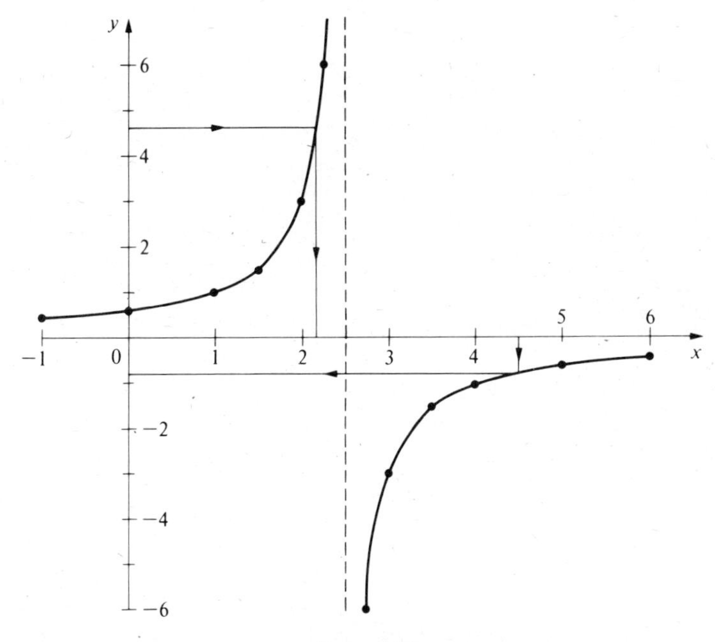

Figure 4.15

Step 9. When $x = 4.5$, from the graph, $y \approx -0.7$. By calculator, $y = -0.75$.
Step 10. When $y = 4.6$, from the graph, $x \approx 2.2$. By calculator, $x = 2.173913$.

3. Draw the graph of $y = 3x^{\frac{1}{2}}$ between $x = 0$ and $x = 10$. Note: y has no real values when x is negative. Consequently no part of the curve lies to the left of the origin.

Step 1.	x	0	1	2	3	4	5	6	7	8	9	10
Step 2.	$x^{\frac{1}{2}}$	0	1	1.41	1.73	2	2.24	2.45	2.65	2.83	3	3.16
Step 3.	$3x^{\frac{1}{2}}$	0	3	4.23	5.19	6	6.72	7.35	7.95	8.49	9	9.48

Step 4. Choose a suitable scale for x.
Step 5. There is no symmetry on this curve.
Step 6. Choose a suitable scale for y. (See Fig. 4.16.)

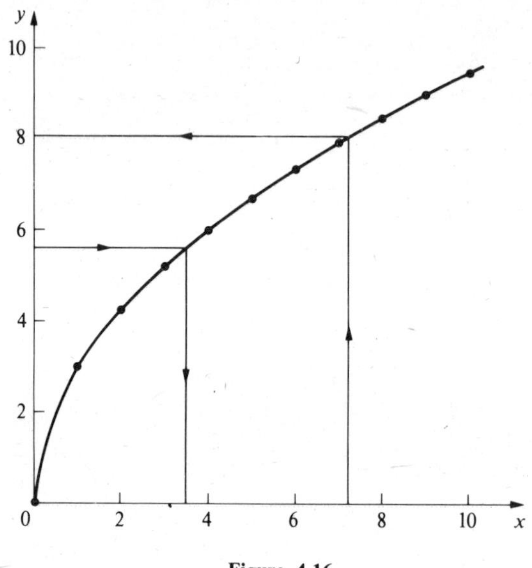

Figure 4.16

Step 7. When $x = 7.2$, from the graph, $y \approx 8.1$. By calculator, $y = 8.0498447$.
Step 8. When $y = 5.6$, from the graph, $x \approx 3.5$. By calculator, $x = 3.4844444$.

Exercise 4.2

Construct suitable tables of values and plot the graphs of the following equations between the indicated values of x. Use the graphs to determine the values of y or x required. Wherever possible check those values by calculator.

1. $y = 4/x$ between $x = -3$ and $x = 3$. Determine y when $x = 2.6$ and x when $y = 8.4$.

2. $y = -1/2x$ between $x = -5$ and $x = 5$. Determine y when $x = -1.4$ and x when $y = -0.9$.
3. $y = \dfrac{48}{2x-1}$ between $x = 2$ and $x = 12$. Determine y when $x = 4.9$ and x when $y = 13.7$
4. $y = \dfrac{100}{3-2x}$ between $x = -3$ and $x = -10$. Determine y when $x = -5.2$ and x when $y = 6.8$.
5. $y = \frac{5}{2}x^{1/2}$ between $x = 0$ and $x = 10$. Determine y when $x = 8.6$ and x when $y = 3.9$.
6. $y = -\frac{5}{8}x^{1/2}$ between $x = 10$ and $x = 100$. Determine y when $x = 69$ and x when $y = 4.7$.
7. $y = 3(x-1)^{\frac{1}{2}}$ between $x = 1$ and $x = 11$. Determine y when $x = 8.2$ and x when $y = 5.6$.

4.3 The conversion of certain curves to straight-line graphs

The one graph which is easy to recognize no matter how little of the graph is visible is the straight-line graph. Consequently it is often of advantage to transform the equation of a curve so that the resulting equation is that of a straight line.

The transformation of $y = ax^2 + b$

By substituting $x^2 = X$ the equation becomes $y = aX + b$. This is the equation of a straight line obtained by plotting y against X. The gradient of the line is a and the intercept is b.

Example

The following table of values of x and y indicates a relationship of the form $y = ax^2 + b$. Calculate the values of a and b. From the graph determine y when $x = 6.3$ and x when $y = 16.4$.

x	1	4	7	10
y	10.5	18	34.5	60

Substitute $x^2 = X$ and construct a new table for X and y:

X	1	16	49	100
y	10.5	18	34.5	60

Fig. 4.17 represents the graph of y plotted against X. The four points lie on a straight line, AB. The line crosses the y axis at $y = 10$. Then $b = 10$. The gradient of $AB = 50/100 = 1/2$. The relationship between y and x is:

$$y = \tfrac{1}{2}x^2 + 10$$

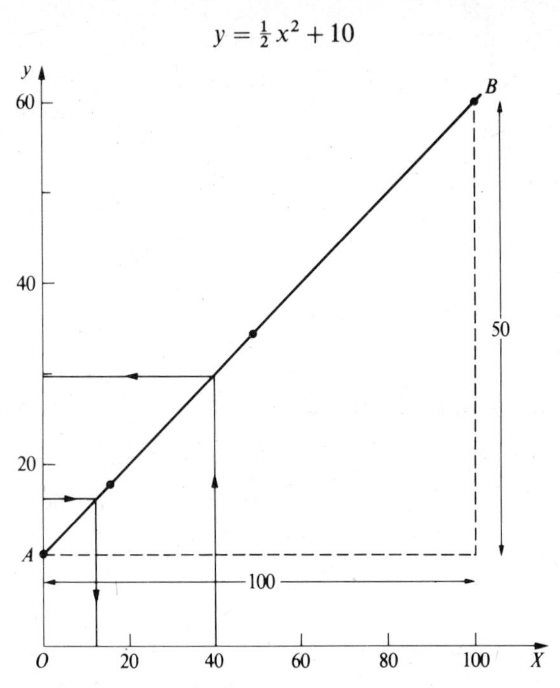

Figure 4.17

When $x = 6.3$, $X = 39.69 \approx 40$. From the graph $y \approx 29.5$. By formula and calculator $y = 29.845$.

When $y = 16.4$, $X \approx 12.5$, $x = \sqrt{12.5} = 3.5$. By the formula and calculator $x = 3.5777088$.

The transformation of $y = p + q/x$

By substituting $1/x = X$ the equation becomes:

$$y = p + qX = qX + p$$

Again this is the equation of a straight line obtained by plotting y against X. The gradient is q and the intercept is p.

Example

The following table of values of x and y refers to a relationship of the form $y = p + q/x$. Calculate the values of p and q. From the graph determine y when $x = \frac{1}{4}$ and x when $y = 35$.

x	1/10	1/5	1/2	1
y	0	25	40	45

Substitute $1/x = X$ and construct a new table for X and y:

X	10	5	2	1
y	0	25	40	45

Fig. 4.18 represents the graph of y plotted against X. The four points plotted lie on a straight line, PQ. The line crosses Oy at $y = 50$. Then $p = 50$. The gradient of $PQ = -50/10 = -5$. Then $q = -5$. The relationship between y and x is:

$$y = 50 - 5/x$$

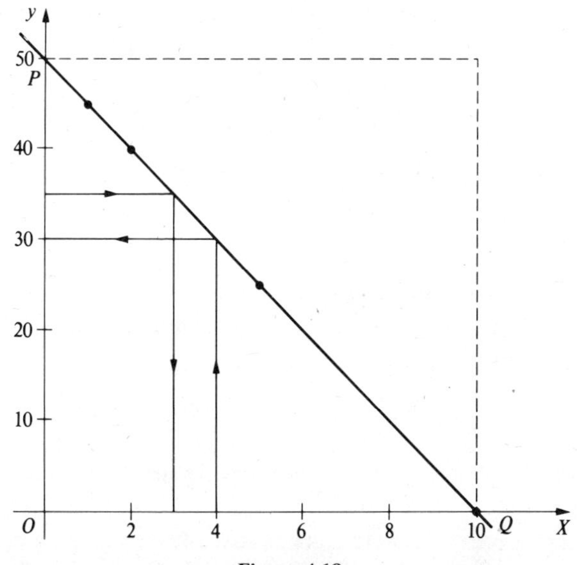

Figure 4.18

When $x = \frac{1}{4}$, $X = 4$. From the graph, $y = 30$. By calculator, $y = 30$.
When $y = 35$, from the graph, $X = 3$, i.e. $x = \frac{1}{3}$. By calculator, $x = \frac{1}{3}$.

Exercise 4.3

Assume that the data in the following tables are consistent with a relationship of the form $y = ax^2 + b$. Make a suitable transformation and plot the revised data on a graph. From the graph calculate the values of a and b.

1.

x	1	3	8	10
y	5.25	7.25	21	30

From the graph determine y when $x = 7.2$ and x when $y = 14.5$.

2.

x	3	5	7	10
y	55.5	47.5	35.5	10

From the graph determine y when $x = 1.6$ and x when $y = 21.9$.

3.

x	2	5	10	11
y	12.4	25	70	82.6

From the graph determine y when $x = 9.2$ and x when $y = 53.5$.

4.

x	5	7	9	10
y	50	2	−62	−100

From the graph determine y when $x = 6.8$ and x when $y = 60$.

5.

x	1/10	1/5	1/2	1
y	20	15	12	11

Assume $y = p + q/x$. Calculate p and q. From the graph determine y when $x = 0.45$ and x when $y = 13.7$.

Assume the data in the following tables are consistent with a relationship of the form $y = p + q/x$. From the graph, obtained after a suitable transformation, calculate the values of p and q.

6.

x	1/5	2/5	4/5	6/5
y	25	20	$17\frac{1}{2}$	$16\frac{2}{3}$

From the graph determine y when $x = 0.5$ and x when $y = 23.2$.

7.

x	1/8	1/6	3/4	1
y	42	44	$48\frac{2}{3}$	49

From the graph determine y when $x = 0.8$ and x when $y = 43.6$.

8.

x	1	4	5	10
y	50	57.5	58	59

From the graph determine y when $x = 7.7$ and x when $y = 53.4$.

4.4 Some standard curves

The parabola is a standard section of a cone. The curve is important in elementary concepts of gravitational motion. Its reflective properties have useful applications in the field of light.

Other curves which are important in many ways have the following standard equations:

$$x^2 + y^2 = a^2$$
$$x^2/a^2 + y^2/b^2 = 1$$
$$x^2/a^2 - y^2/b^2 = 1$$
$$xy = c^2$$

The curve $x^2 + y^2 = a^2$

Fig. 4.19 is a sketch of the curve. It is a circle, centre O, radius a.

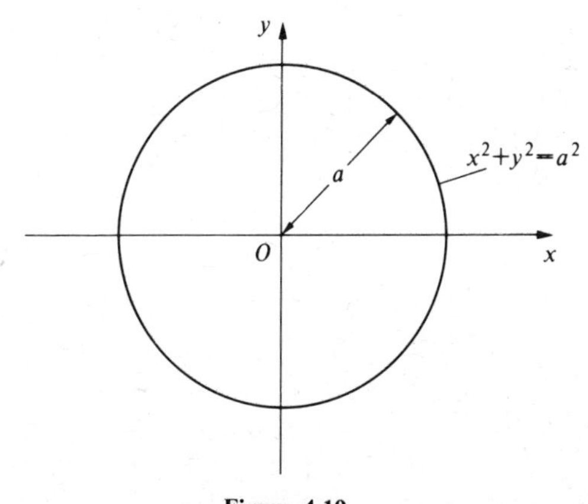

Figure 4.19

It is symmetrical about Ox (detected in the equation because y occurs only as an even power). It is symmetrical about Oy (x occurs only as an even power). It is symmetrical about the axis through O perpendicular to its plane.

The curve $x^2/a^2 + y^2/b^2 = 1$

Fig. 4.20 is a sketch of the curve. It is an ellipse, centre O. Its axes AA^1 and BB^1 are $2a$ and $2b$ respectively. It is symmetrical about Ox. It is symmetrical about Oy. It is symmetrical about the axis through O perpendicular to its plane. When a circle is drawn on AA^1 as diameter then the ratio QN/PN is always a/b for all positions of P on the ellipse.

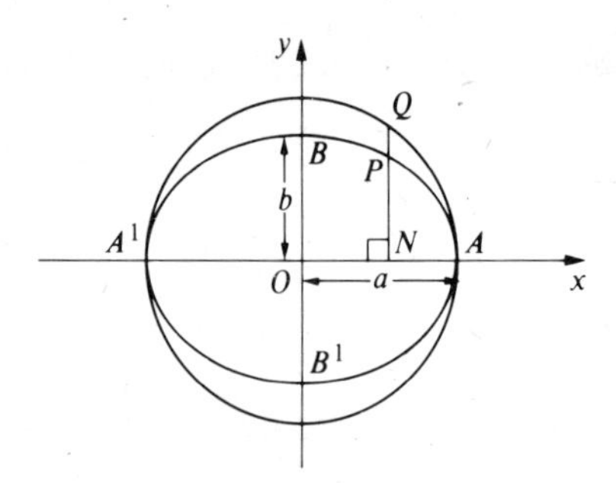

Figure 4.20

The curve $x^2/a^2 - y^2/b^2 = 1$

Fig. 4.21 represents a sketch of the curve. It is a hyperbola with centre O. The curve lies entirely within the angle between two lines, HK and LM. They are the asymptotes. Their gradients are b/a and $-b/a$ respectively. The curve is

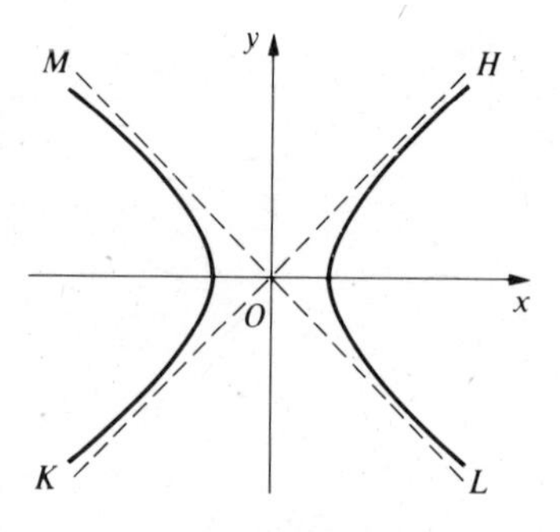

Figure 4.21

symmetrical about Ox. The curve is symmetrical about Oy. The curve is symmetrical about the axis through O perpendicular to its plane.

The curve $xy = c^2$

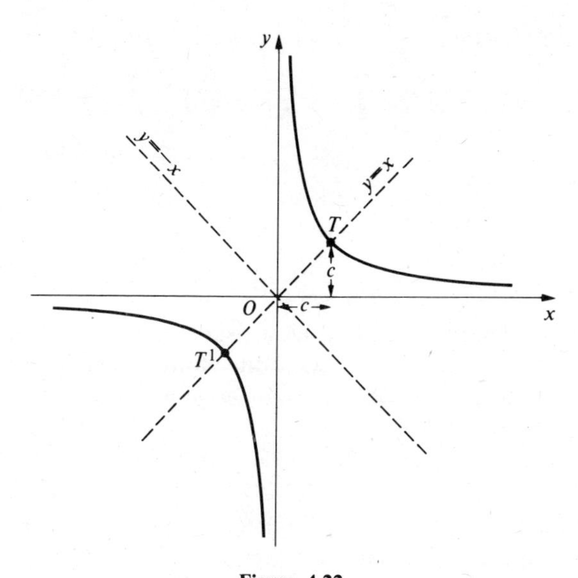

Figure 4.22

Fig. 4.22 represents a sketch of the curve. It is a rectangular hyperbola with centre O. The asymptotes are Ox and Oy. The hyperbola is called rectangular because the asymptotes are perpendicular to each other. The co-ordinates of T, T^1 are (c, c) and $(-c, -c)$ respectively. The curve is not symmetrical about Ox or Oy. It is symmetrical about the bisectors of the angles between Ox and Oy. It is also symmetrical about the axis through O perpendicular to its plane.

5

The Use of Graphs to Determine Values and Established Power Laws

Chapter 2 of *Mathematics Level 2* dealt with graphs of straight lines and the method of determining linear laws from plotted data. Often, though, experimental data and sets of observed measurements are not related by linear laws but by a law which is often called the power law. This is of the form

$$y = ax^n \qquad 5.1$$

where x and y represent the pairs of related values. The values of a and n depend on the nature of the source of the data, e.g. the kind of machine, or the particular technological or scientific principle.

Usually the object in drawing the graph is to use it to determine the values of a and n as well as to use the graph to determine other related values of x and y. If y were plotted against x the shape of the curve would depend on the values of a and n and also on the relative scales of the two axes. The task of recognizing the particular shape which corresponds to particular values of a and n would be almost impossible.

The one shape which is easily recognized is that of the straight line. To achieve this shape of a straight-line graph we carry out a transformation of *5.1*.

5.1 The use of logarithms to reduce laws of the type $y = ax^n$ to straight line form

Take logs to base 10 of *5.1* to obtain:

$$\log y = \log a + \log x^n$$
$$= \log a + n \log x$$
$$\log y = n \log x + \log a \qquad 5.2$$

Put $\log x = X$ and $\log y = Y$. Then *5.2* becomes:

$$Y = nX + \log a \qquad 5.3$$

By plotting Y against X, *5.3* represents a straight-line graph. Its gradient is n; the intercept is $\log a$. Consequently, by determining the gradient of the line, we obtain n, and by determining the intercept of the line we obtain $\log a$. Fig. 5.1 represents a sketch of such a graph. From it:

$$OA = \log a \qquad 5.4$$

$$a = 10^{OA} \qquad 5.5$$

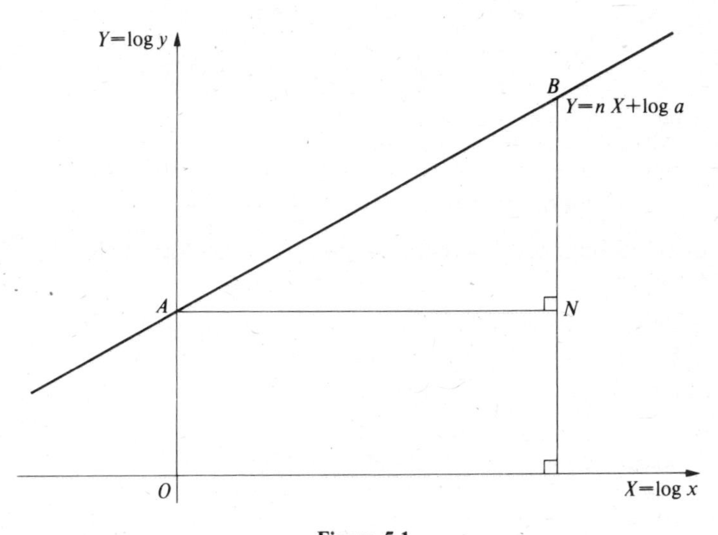

Figure 5.1

Suppose B is a point on the line and that AN and NB are parallel to OX and OY respectively such that AN is a convenient integral value. Then:

$$n = BN/AN \qquad 5.6$$

Particular examples of the power law are:

$s = \frac{1}{2}at^2$, relating distance and time when a is constant
$E = \frac{1}{2}mv^2$, relating kinetic energy and velocity of a given body
$T = 2\pi\sqrt{\frac{L}{g}}$, relating time of oscillation of a pendulum and its length
$p = c/v, = cv^{-1}$, relating pressure and volume of a gas

Examples

1. The energy of a moving body varies with the velocity in the following manner:

Velocity (v) m/s	10	15	20	25	30
Energy (E) joules	2500	5625	10 000	15 625	22 500

Assume:

$$E = a \, . \, v^n$$

Take logs:

$$\log E = \log a + \log v^n$$
$$= \log a + n \log v$$

Now construct a new table:

$\log v$	1	1.1761	1.3010	1.3979	1.4771
$\log E$	3.3979	3.7501	4.0000	4.1939	4.3522

By correcting these figures to two decimal places the graph in Fig. 5.2 is obtained.

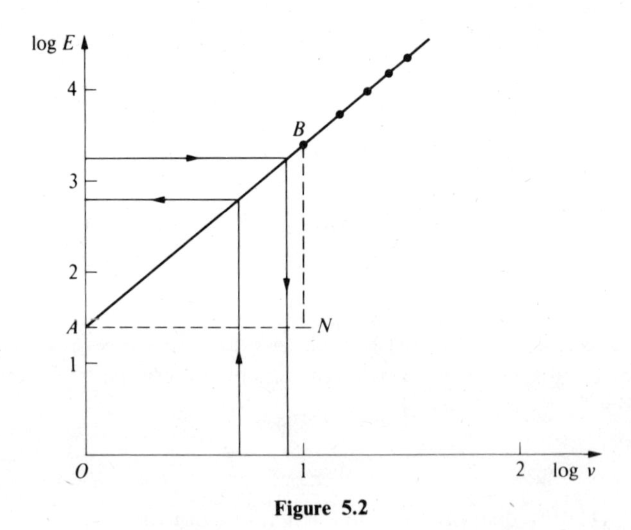

Figure 5.2

From the graph, $\log a = 1.4$; $a = 10^{1.4} = 25.118864 \approx 25.12$ $n = BN/AN = (3.4 - 1.4)/1 = 2$. The law appears to be $E = 25.12\, v^2$

From the graph, when $v = 5$, $\log v = 0.6987$ (by calculator), i.e. lo

$v \approx 0.70$. Then $\log E = 2.75$; by calculator, $E = 10^{2.75} = 562.34 \approx 562$.

When $E = 1778$, $\log E = 3.25$. Then $\log v = 0.93$ from the graph, $v = 8.5113804$ (by calculator) ≈ 8.5.

As a check on the graphical method the following is a calculation to obtain the values of a and n. Substituting in $\log E = \log a + n \log v$ from the original table of values:

$$\log 22\,500 = \log a + n \log 30 \qquad \text{(i)}$$

$$\log 2500 = \log a + n \log 10 \qquad \text{(ii)}$$

(i) – (ii):

$$\log 22\,500 - \log 2500 = n \log 30 - n \log 10$$

$$\log \frac{22\,500}{2500} = n \cdot \log \frac{30}{10}$$

$$\log 9 = n \log 3$$

$$2 \log 3 = n \log 3, \; n = 2$$

Substitute for n in (ii):

$$\log 2500 = \log a + 2 \log 10$$

$$\log 2500 - \log 100 = \log a$$

$$\log 25 = \log a$$

$$a = 25$$

2. The length of the side of a square sheet of metal plate of uniform thickness and density varies with the mass of the plate. The following table of values relates s, the side of the plate in millimetres, to W, the mass of the plate in kilograms.

W (kg)	4	9	16	25	36
s (mm)	220	330	440	550	660

Suppose $s = a \, . \, W^n$. Then:

$$\log s = \log a + n \log W$$

$$\log s = n \log W + \log a$$

The following table relates $\log s$ and $\log W$:

$\log W$	0.6021	0.9542	1.2041	1.3979	1.5563
$\log s$	2.3424	2.5185	2.6435	2.7404	2.8195

From the graph in Fig. 5.3, $\log a \approx 2.04$; $a = 10^{2.04} = 109.64 \approx 110$. Also, from the graph, $n = BN/AN = (3.04 - 2.04)/2 = 1/2$. Therefore $s = 110 \, . \, W^{\frac{1}{2}} = 110 \sqrt{W}$.

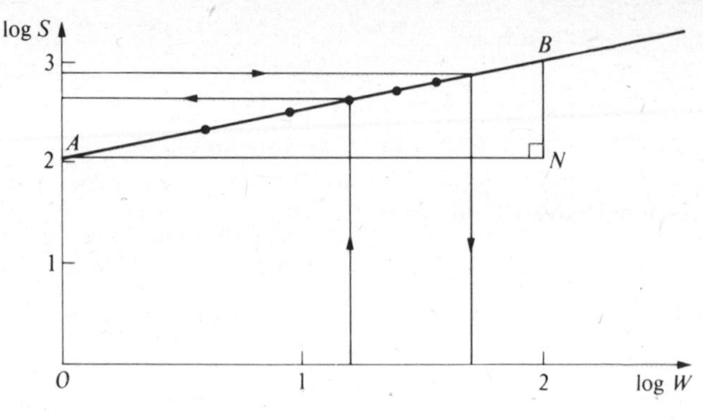

Figure 5.3

From the graph, when $\log W = 1.1$, i.e. $W = 12.589254 \approx 12.6$, then $\log s = 2.58$, $s = 10^{2.58} = 380.1894 \approx 380$.

From the graph, when $\log s = 2.88$, i.e. $s = 10^{2.88} = 758.7758 \approx 759$, then $\log W = 2.65$, $W = 10^{2.65} = 446.68359 \approx 447$.

3. A constant potential difference is applied to the terminals of a variable conductor. The current through the conductor and its resistance are connected by a relation represented by the values of R, in ohms, and I, in amperes, in the following table:

R (ohms)	50	100	150	200	250
I (amperes)	5.00	2.50	1.67	1.25	1.00

Assume $I = a \,.\, R^n$. Then:

$$\log I = \log a + n \log R$$

The new table, for $\log R$ and $\log I$, is:

$\log R$	1.70	2.00	2.18	2.30	2.40
$\log I$	0.70	0.40	0.22	0.10	0.00

From the graph in Fig. 5.4, $\log a = 2.40$, $a = 10^{2.4} = 251.18864 \approx 251$ or, rounded off even more, 250. From the graph, $n = BN/AN = (0.4 - 2.4)/2 = -1$. Therefore $I = 250\,R^{-1}$.

From the graph, when $\log R = 1.25$, $R = 17.782794 \approx 17.8$, then $\log I = 1.15$, $I = 14.125375 \approx 14.13$.

When $\log I = 1.75$, $I = 56.234133 \approx 56.23$, then $\log R = 0.65$, $R = 10^{0.65} = 4.4668359 \approx 4.47$.

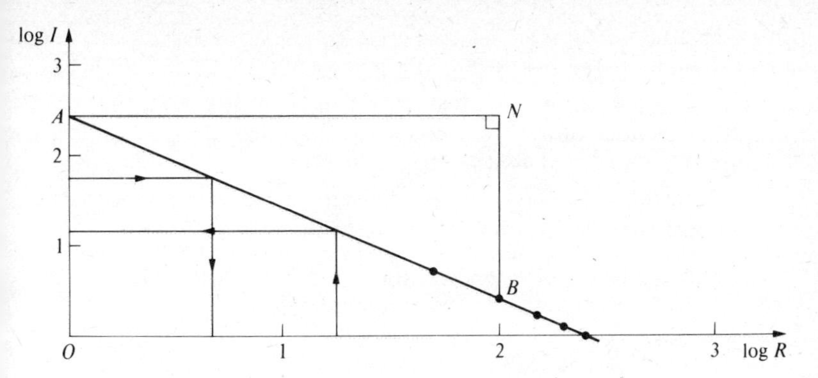

Figure 5.4

Exercise 5.1

1. Draw a graph representing the law $y = 10.x^3$ so as to obtain a straight-line graph. Construct a table by taking the values of x to be 10, 20, 30, 40. Hint: rearrange the equation to $\log y = \log 10 + 3 \log x$. From the graph check that the gradient is 3 and that the intercept is log 10. Determine y when $x = 24$ and x when $y = 12\,000$.
2. Draw a straight-line graph to represent the law $y = 10.x^{-2}$, using the following values of x: 1, 0.8, 0.6, 0.5. From the graph determine y when $x = 0.7$ and x when $y = 200$.
3. Draw a straight-line graph to represent $y = 6.25x^{1.6}$ using the following values of x: 1, 4, 8, 10. From the graph determine y when $x = 5.2$ and x when $y = 63.7$.
4. The masses of circular metal plates vary with the radii of the plates as follows:

Radius (R) mm	200	300	400	500	600
Mass (W) kg	5.00	11.25	20.00	31.25	45.00

   Construct a new table relating $\log R$ to $\log W$. Use it to plot a graph. From the graph verify that the law relating W to R is of the form $W = k.R^2$. Find an approximate value of k. From the graph determine the value of W when $R = 340$ and the value of R when $W = 33.60$.
5. The velocities which a lead weight attains after falling through varying heights are given in the following table:

Height (h) m	16	25	49	64	100
Velocity (V) m/s	5.6	7.0	9.8	11.2	14.0

Construct a new table relating log V to log h. From it draw a graph. From the graph verify that the law relating V to h is of the form $V = a.h^{\frac{1}{2}}$, where a is a constant. Calculate as accurately as you are able the value of a. From the graph determine the value of V when $h = 37$ and the value of h when $V = 12.3$.

6. The current, I amperes, through a given conductor to which a variable voltage is applied, varies according to the following table:

Voltage (V) volts	50	100	150	200	250	300
Current (I) amperes	0.2	0.4	0.6	0.8	1.0	1.2

Determine the law connecting I and V. When $V = 60.5$ determine I and when $I = 0.65$ determine V.

7. The time of oscillation of a simple pendulum and its length are related by a certain law. The following table gives pairs of values of T and L which are related to one another, where T represents the time of oscillation and L the length of the pendulum.

Length (L) m	1	4	9	16	25
Time (T) s	2	4	6	8	10

Determine the law relating T to L. From your graph determine the value of L when T is 5 and the value of T when L is 2.5.

8. A body subjected to different velocities is timed to cover a given distance at a constant speed. The following pairs of values are recorded:

Velocity (V) m/s	5	10	15	20	25
Time (t) s	8.4	4.2	2.8	2.1	1.68

Determine graphically the relation between t and V. From the graph determine the value of t when $V = 22.5$ and the value of V when $t = 3.0$.

9. A curve with an equation of the form $y = ax^n$ is satisfied by the following data:

x	0.42	0.89	2.08
y	6.2	8.5	12.2

Determine a and n. Calculate y when $x = 1.86$ and x when $y = 7.3$.

10. The following table represents recommended cutting speeds for boring holes of diameter d millimetres in a certain metal. Determine the law relating N to d where N is the speed in rev/min, assuming that it is of the form $N = a.d^n$.

d (mm)	20	25	30	35	40
N (rev/min)	336	265	218	184	160

Calculate N when $d = 25.4$ and 38.1 and d when $N = 260$ and 175.

11. The following table represents the relation between the attractive force, F, of two magnetic poles and their distance apart, d.

d	2	4	6	8
F	49.1	12.7	5.5	3.2

Assume $F = a.d^n$. Determine the values of a and n. Determine F when $d = 5$ and 7 and d when $F = 31.6$ and 5.2.

12. The following data relate the pressure and volume of a gas during a certain kind of expansion:

Volume (v) litre	14.35	11.3	9.55	7.70
Pressure (P) kilopascals	2848.2	3595.6	4221.8	5191.4

Determine the law relating P with v. Calculate from your graph the value of P when $v = 12.2$ and 8.60 and the value of v when $P = 3100.7$ and 4918.7.

13. The following data satisfy a relation of the form $W = a.V^n$, where a and n are constants.

V	2.6	5.8	10.9
W	0.0295858	0.0026651	0.0004015

Calculate a and n. Determine W when $V = 4.7$ and V when $W = 0.00123$.

6

Matrices and Determinants

In *Mathematics Level 1* two methods of solution of simultaneous linear equations in two unknowns were adopted: (a) by elimination, and (b) by substitution. By looking at such problems in a different way a new kind of number emerges: a matrix. As a first step towards that end we look again at a simple, familiar problem.

Examples

1. Solve $3x = 5$.
   Step 1. $\frac{1}{3} \times (3x) = \frac{1}{3} \times 5$
   Step 2. $(\frac{1}{3} \times 3) \times x = 5/3$, the associative law
   Step 3. $1 \times x = 5/3$ ($\frac{1}{3}$ and 3 are inverses under multiplication)
   Step 4. $x = 5/3$
2. Solve $ax = b$, where a and b are constants $\neq 0$.
   Step 1. $\frac{1}{a} \times (ax) = \frac{1}{a} \times b$, or $a^{-1} \times (ax) = a^{-1} . b$
   Step 2. $\left(\frac{1}{a} \times a\right) \times x = \frac{b}{a}$, or $(a^{-1} . a) . x = a^{-1} . b$
   Step 3. $1 \times x = b/a$, or $x = a^{-1} . b$
   Step 4. $x = b/a$
3. Solve
$$\left.\begin{aligned} 2x + y &= 5 \\ x + y &= 3 \end{aligned}\right\}$$
   Rewrite:
$$\begin{pmatrix} 2 & 1 \\ 1 & 1 \end{pmatrix} . \begin{pmatrix} x \\ y \end{pmatrix} = \begin{pmatrix} 5 \\ 3 \end{pmatrix} \qquad 6.1$$

Put $\begin{pmatrix} 2 & 1 \\ 1 & 1 \end{pmatrix} = A, \quad \begin{pmatrix} x \\ y \end{pmatrix} = X, \quad \begin{pmatrix} 5 \\ 3 \end{pmatrix} = B$

Equation *6.1* becomes $A.X = B$, now an equation like those in Examples (1) and (2) above. To solve this equation we would require:

$$A^{-1}.(A.X) = A^{-1}.B$$
$$X = A^{-1}.B$$

For this to be possible we need to give meanings to:

$$\begin{pmatrix} 2 & 1 \\ 1 & 1 \end{pmatrix} \begin{pmatrix} x \\ y \end{pmatrix} \begin{pmatrix} 5 \\ 3 \end{pmatrix} \text{ and } A^{-1} \text{ when } A = \begin{pmatrix} 2 & 1 \\ 1 & 1 \end{pmatrix}$$

They are all examples of matrices. The solution of the above equations is left until the next chapter.

Definition

A matrix is a rectangular array of numbers arranged in rows and columns. When there are m rows and n columns it is called an $m \times n$ matrix.

For the present we shall restrict ourselves to the study of 2×2 matrices. Examples of 2×2 matrices are:

$$\begin{pmatrix} 1 & 3 \\ 2 & 5 \end{pmatrix} \begin{pmatrix} 4 & -1 \\ 7 & -2 \end{pmatrix} \begin{pmatrix} -1 & -4 \\ -5 & -2 \end{pmatrix} \begin{pmatrix} 1 & 0 \\ 0 & 1 \end{pmatrix} \begin{pmatrix} 2 & 0 \\ 0 & 0 \end{pmatrix} \begin{pmatrix} 0 & -3 \\ 2 & 0 \end{pmatrix}$$

6.1 The sum and difference of two 2×2 matrices

Definitions

The sum of two 2×2 matrices is defined by the following rule:

$$\begin{pmatrix} a & b \\ c & d \end{pmatrix} + \begin{pmatrix} A & B \\ C & D \end{pmatrix} = \begin{pmatrix} a+A & b+B \\ c+C & d+D \end{pmatrix} \qquad 6.2$$

The difference between two matrices is defined by the rule:

$$\begin{pmatrix} A & B \\ C & D \end{pmatrix} - \begin{pmatrix} a & b \\ c & d \end{pmatrix} = \begin{pmatrix} A-a & B-b \\ C-c & D-d \end{pmatrix} \qquad 6.3$$

In other words, the sum of two matrices is obtained by adding together

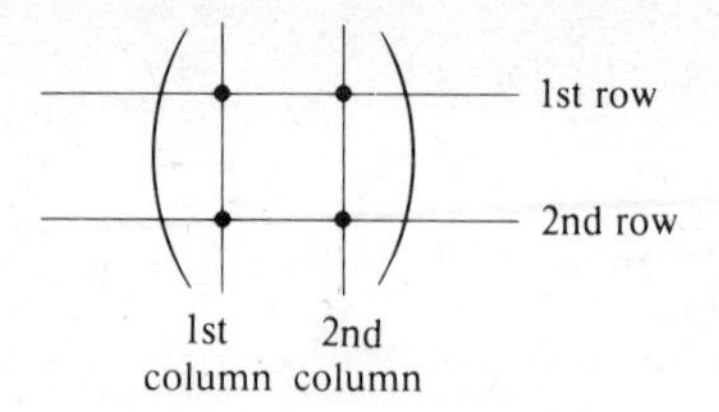

Figure 6.1

corresponding elements of the matrices. The difference between two matrices is obtained by finding the differences between corresponding elements. The term 'corresponding elements' needs clarification. Fig. 6.1 represents a 2×2 matrix in which the dots represent the positions of the four elements. The elements in the matrices which are in the first row and the first column are corresponding elements. The elements in the first row and the second column are corresponding elements, and so on.

$$\text{When } X = \begin{pmatrix} a_{11} & a_{12} \\ a_{21} & a_{22} \end{pmatrix} \text{ and } Y = \begin{pmatrix} b_{11} & b_{12} \\ b_{21} & b_{22} \end{pmatrix}$$

$$\text{then } X + Y = \begin{pmatrix} a_{11} + b_{11} & a_{12} + b_{12} \\ a_{21} + b_{21} & a_{22} + b_{22} \end{pmatrix}$$

The suffixes of the elements in X and Y correspond to the row and column positions of the elements in the matrices: the first suffix for the row and the second suffix for the column.

Examples

1. $\begin{pmatrix} 3 & 2 \\ 1 & 4 \end{pmatrix} + \begin{pmatrix} 9 & 5 \\ 2 & 6 \end{pmatrix} = \begin{pmatrix} 3+9 & 2+5 \\ 1+2 & 4+6 \end{pmatrix} = \begin{pmatrix} 12 & 7 \\ 3 & 10 \end{pmatrix}$

2. $\begin{pmatrix} 4 & 0 \\ -5 & 7 \end{pmatrix} + \begin{pmatrix} -3 & 8 \\ 6 & 2 \end{pmatrix} = \begin{pmatrix} 4-3 & 0+8 \\ -5+6 & 7+2 \end{pmatrix} = \begin{pmatrix} 1 & 8 \\ 1 & 9 \end{pmatrix}$

3. $\begin{pmatrix} 5a & -3a \\ -2a & 4a \end{pmatrix} - \begin{pmatrix} 3a & 4a \\ -4a & -7a \end{pmatrix} = \begin{pmatrix} 8a & a \\ -6a & -3a \end{pmatrix}$

4. $\begin{pmatrix} 2 & 3 \\ 4 & 1 \end{pmatrix} - \begin{pmatrix} 1 & 0 \\ 2 & 0 \end{pmatrix} = \begin{pmatrix} 2-1 & 3-0 \\ 4-2 & 1-0 \end{pmatrix} = \begin{pmatrix} 1 & 3 \\ 2 & 1 \end{pmatrix}$

5. $\begin{pmatrix} a+2b & 6b \\ 3a-4b & 5a \end{pmatrix} + \begin{pmatrix} 2a+3b & a-4b \\ 6a+3b & a+2b \end{pmatrix} = \begin{pmatrix} 3a+5b & a+2b \\ 9a-b & 6a+2b \end{pmatrix}$

Exercise 6.1

Express as a single matrix each of the following:

1. $\begin{pmatrix} 1 & 2 \\ 3 & 4 \end{pmatrix} + \begin{pmatrix} 2 & 4 \\ 3 & 1 \end{pmatrix}$

2. $\begin{pmatrix} 6 & 11 \\ 7 & 4 \end{pmatrix} + \begin{pmatrix} 9 & 13 \\ 14 & 2 \end{pmatrix}$

3. $\begin{pmatrix} 9 & 0 \\ 0 & 8 \end{pmatrix} + \begin{pmatrix} 0 & 7 \\ 5 & 0 \end{pmatrix}$

4. $\begin{pmatrix} 11 & -1 \\ -7 & 4 \end{pmatrix} + \begin{pmatrix} -3 & 15 \\ 12 & -6 \end{pmatrix}$

5. $\begin{pmatrix} 3a & 2a \\ 5a & 6a \end{pmatrix} + \begin{pmatrix} 0 & 9a \\ 4a & 11a \end{pmatrix}$

6. $\begin{pmatrix} -2b & 3b \\ 4b & -7b \end{pmatrix} + \begin{pmatrix} 8b & 2b \\ -6b & -9b \end{pmatrix}$

7. $\begin{pmatrix} a+b & a-b \\ a+2b & b-2a \end{pmatrix} + \begin{pmatrix} a-2b & 2a+3b \\ 4a-3b & 3a+4b \end{pmatrix}$

8. $\begin{pmatrix} 2 & 4 \\ 3 & 1 \end{pmatrix} - \begin{pmatrix} 1 & 3 \\ 2 & 0 \end{pmatrix}$

9. $\begin{pmatrix} 15 & 12 \\ 17 & 23 \end{pmatrix} - \begin{pmatrix} 9 & 11 \\ 14 & 17 \end{pmatrix}$

10. $\begin{pmatrix} 12 & 0 \\ -13 & 2 \end{pmatrix} - \begin{pmatrix} -7 & 5 \\ 0 & 15 \end{pmatrix}$

11. $\begin{pmatrix} 35 & 41 \\ -62 & 53 \end{pmatrix} - \begin{pmatrix} -27 & 34 \\ -93 & -27 \end{pmatrix}$

12. $\begin{pmatrix} 73 & -26 \\ 48 & -53 \end{pmatrix} - \begin{pmatrix} 0 & 28 \\ 36 & 19 \end{pmatrix}$

13. $\begin{pmatrix} 3a & 8a \\ -7a & 15a \end{pmatrix} - \begin{pmatrix} -8a & -3a \\ 15a & -7a \end{pmatrix}$

14. $\begin{pmatrix} 3a+2b & 0 \\ 7a-5b & 3b-2a \end{pmatrix} - \begin{pmatrix} 4a-7b & 2a-5b \\ 3a-4b & 6b+3a \end{pmatrix}$

Examples

1. Write $2\cdot\begin{pmatrix}2 & 1\\ 4 & -3\end{pmatrix}$ as a 2×2 matrix.

   The rule for ordinary numbers is $2x = x + x$, so here, with matrices:

$$2\cdot\begin{pmatrix}2 & 1\\ 4 & -3\end{pmatrix}=\begin{pmatrix}2 & 1\\ 4 & -3\end{pmatrix}+\begin{pmatrix}2 & 1\\ 4 & -3\end{pmatrix}=\begin{pmatrix}4 & 2\\ 8 & -6\end{pmatrix}\text{ by } 6.2$$

$$=\begin{pmatrix}2\times 2 & 2\times 1\\ 2\times 4 & 2\times -3\end{pmatrix}$$

2. $$3\begin{pmatrix}a & b\\ c & d\end{pmatrix}=\begin{pmatrix}a & b\\ c & d\end{pmatrix}+\begin{pmatrix}a & b\\ c & d\end{pmatrix}+\begin{pmatrix}a & b\\ c & d\end{pmatrix}$$

$$=\begin{pmatrix}3a & 3b\\ 3c & 3d\end{pmatrix}$$

3. Generally: $$p\begin{pmatrix}a & b\\ c & d\end{pmatrix}=\begin{pmatrix}pa & pb\\ pc & pd\end{pmatrix} \qquad 6.4$$

4. $$p.\begin{pmatrix}a & b\\ c & d\end{pmatrix}+q.\begin{pmatrix}x & y\\ z & t\end{pmatrix}=\begin{pmatrix}pa & pb\\ pc & pd\end{pmatrix}+\begin{pmatrix}qx & qy\\ qz & qt\end{pmatrix}$$

$$=\begin{pmatrix}pa+qx & pb+qy\\ pc+qz & pd+qt\end{pmatrix}$$

Exercise 6.2

Simplify:

1. $2.\begin{pmatrix}2 & 1\\ 3 & 4\end{pmatrix}+3.\begin{pmatrix}1 & 2\\ 4 & 3\end{pmatrix}$

2. $3.\begin{pmatrix}6 & 1\\ -2 & 3\end{pmatrix}+5.\begin{pmatrix}-4 & 7\\ 0 & 9\end{pmatrix}$

3. $6.\begin{pmatrix}1 & 0\\ 0 & 1\end{pmatrix}+7.\begin{pmatrix}1 & 0\\ 0 & -1\end{pmatrix}-4.\begin{pmatrix}0 & 1\\ -1 & 0\end{pmatrix}$

4. $2.\begin{pmatrix}a & b\\ c & d\end{pmatrix}+3.\begin{pmatrix}a & -b\\ -c & d\end{pmatrix}$

5. $p.\begin{pmatrix} a & 0 \\ 0 & a \end{pmatrix} + q.\begin{pmatrix} 0 & -a \\ a & 0 \end{pmatrix} + r.\begin{pmatrix} 1 & 1 \\ 1 & 0 \end{pmatrix}$

6. $3.\begin{pmatrix} ax & 0 \\ 0 & by \end{pmatrix} + 7.\begin{pmatrix} 0 & bx \\ -ay & 0 \end{pmatrix}$

7. $\frac{1}{2}.\begin{pmatrix} 2 & 0 \\ 0 & 2 \end{pmatrix}$

8. $\frac{1}{3}.\begin{pmatrix} 3 & 0 \\ 0 & 3 \end{pmatrix}$

9. $\frac{1}{a}.\begin{pmatrix} a & 0 \\ 0 & a \end{pmatrix}$

10. $\frac{1}{2a}.\begin{pmatrix} a & 0 \\ 0 & a \end{pmatrix} + \frac{1}{2b}.\begin{pmatrix} b & 0 \\ 0 & b \end{pmatrix}$

6.2 The product of two matrices

In Example (3) at the beginning of the chapter:

$$\left.\begin{aligned} 2x + y &= 5 \\ x + y &= 3 \end{aligned}\right\} \text{ was replaced by } \begin{pmatrix} 2 & 1 \\ 1 & 1 \end{pmatrix} \cdot \begin{pmatrix} x \\ y \end{pmatrix} = \begin{pmatrix} 5 \\ 3 \end{pmatrix}$$

The operation between the matrices on the LHS is taken to be multiplication. In order that the two statements agree the following statement must be true:

$$\begin{pmatrix} 2 & 1 \\ 1 & 1 \end{pmatrix} \cdot \begin{pmatrix} x \\ y \end{pmatrix} = \begin{pmatrix} 2x + y \\ x + y \end{pmatrix}$$

That is, a 2×2 matrix times a 2×1 matrix $=$ a 2×1 matrix. More generally, it would mean:

$$\begin{pmatrix} \underline{a} & \underline{b} \\ \underline{c} & \underline{d} \end{pmatrix} \cdot \begin{pmatrix} \underline{x} \\ \underline{y} \end{pmatrix} = \begin{pmatrix} \underline{ax} + \underline{by} \\ \underline{cx} + \underline{dy} \end{pmatrix} \qquad 6.5$$

$$\text{or} \quad A.X = Z$$

To obtain any element of Z multiply the elements in the corresponding row of A by the elements in X. The elements underlined similarly are the ones which are multiplied together. The resulting products are added.

Examples

Express as single matrices:

1. $\begin{pmatrix} 3 & 2 \\ 4 & 1 \end{pmatrix} \cdot \begin{pmatrix} 1 \\ 2 \end{pmatrix}$

   The product $= \begin{pmatrix} 3 \times 1 + 2 \times 2 \\ 4 \times 1 + 1 \times 2 \end{pmatrix} = \begin{pmatrix} 3 + 4 \\ 4 + 2 \end{pmatrix} = \begin{pmatrix} 7 \\ 6 \end{pmatrix}$

2. $\begin{pmatrix} 6 & -5 \\ -2 & 3 \end{pmatrix} \cdot \begin{pmatrix} 4 \\ -3 \end{pmatrix}$

   The product $= \begin{pmatrix} 6 \times 4 + -5 \times -3 \\ -2 \times 4 + 3 \times -3 \end{pmatrix} = \begin{pmatrix} 39 \\ -17 \end{pmatrix}$

3. $\begin{pmatrix} a & b \\ -b & a \end{pmatrix} \cdot \begin{pmatrix} 1 \\ -1 \end{pmatrix}$

   The product $= \begin{pmatrix} a - b \\ -b - a \end{pmatrix} = \begin{pmatrix} a - b \\ -(a + b) \end{pmatrix}$

4. $\begin{pmatrix} a & b \\ -b & -a \end{pmatrix} \cdot \begin{pmatrix} b \\ a \end{pmatrix}$

   The product $= \begin{pmatrix} ab + ba \\ -b^2 - a^2 \end{pmatrix} = \begin{pmatrix} 2ab \\ -(a^2 + b^2) \end{pmatrix}$

Exercise 6.3

Express as single matrices:

1. $\begin{pmatrix} 1 & 2 \\ 3 & 4 \end{pmatrix} \cdot \begin{pmatrix} 2 \\ 1 \end{pmatrix}$

2. $\begin{pmatrix} 3 & 4 \\ -2 & 1 \end{pmatrix} \cdot \begin{pmatrix} 1 \\ 5 \end{pmatrix}$

3. $\begin{pmatrix} 6 & -4 \\ 3 & 2 \end{pmatrix} \cdot \begin{pmatrix} -1 \\ 2 \end{pmatrix}$

4. $\begin{pmatrix} -3 & 0 \\ 5 & 2 \end{pmatrix} \cdot \begin{pmatrix} -3 \\ -7 \end{pmatrix}$

5. $\begin{pmatrix} a & b \\ b & a \end{pmatrix} \cdot \begin{pmatrix} 1 \\ 0 \end{pmatrix}$

. $\begin{pmatrix} a & 2a \\ -a & 2a \end{pmatrix} \cdot \begin{pmatrix} 1 \\ -1 \end{pmatrix}$

. $\begin{pmatrix} 3a & 4b \\ 5a & -7b \end{pmatrix} \cdot \begin{pmatrix} -a \\ b \end{pmatrix}$

. $\begin{pmatrix} a & b \\ c & d \end{pmatrix} \cdot \begin{pmatrix} A \\ B \end{pmatrix}$

. $\begin{pmatrix} a & b \\ c & d \end{pmatrix} \cdot \begin{pmatrix} C \\ D \end{pmatrix}$

efinition of the product of two 2 × 2 matrices

$$\begin{pmatrix} a & b \\ c & d \end{pmatrix} \cdot \begin{pmatrix} A & C \\ B & D \end{pmatrix} = \begin{pmatrix} (aA+bB) & (aC+bD) \\ (cA+dB) & (cC+dD) \end{pmatrix} \qquad 6.6$$

he schematic diagram in Fig. 6.2 is helpful to determine such a product.

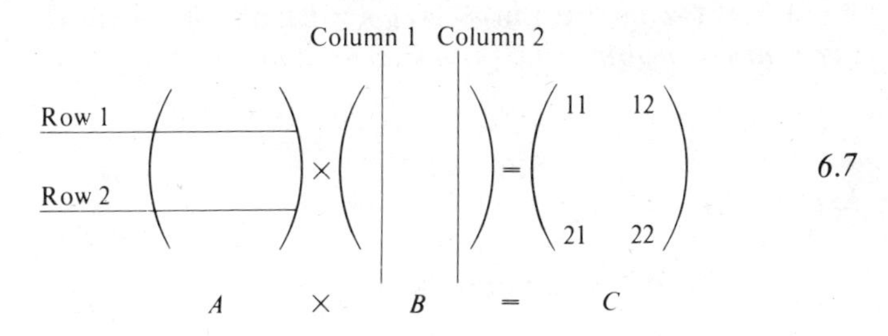

Figure 6.2

tem 1. The elements in a row of A are noted.
tem 2. The elements in a column of B are noted.
tem 3. The position of each element in C is determined by two numbers: ne first represents a row of A, the second a column of B.
tem 4. Each element of C is calculated by multiplying together the lements of a row of A by the corresponding elements of a column of B. The ndividual products are added together.

xamples

. Calculate $\begin{pmatrix} 1 & 2 \\ 3 & 4 \end{pmatrix} \cdot \begin{pmatrix} 5 & 7 \\ 6 & 8 \end{pmatrix}$

$$\text{The product} = \begin{pmatrix} (1\times5+2\times6) & (1\times7+2\times8) \\ (3\times5+4\times6) & (3\times7+4\times8) \end{pmatrix}$$

$$= \begin{pmatrix} (5+12) & (7+16) \\ (15+24) & (21+32) \end{pmatrix} = \begin{pmatrix} 17 & 23 \\ 39 & 53 \end{pmatrix}$$

2. Calculate $\begin{pmatrix} 5 & -2 \\ -9 & 4 \end{pmatrix} \cdot \begin{pmatrix} 3 & 2 \\ -7 & -6 \end{pmatrix}$

$$\text{The product} = \begin{pmatrix} (15+14) & (10+12) \\ (-27-28) & (-18-24) \end{pmatrix} = \begin{pmatrix} 29 & 22 \\ -55 & -42 \end{pmatrix}$$

3. Calculate $\begin{pmatrix} 3 & 2 \\ -7 & -6 \end{pmatrix} \cdot \begin{pmatrix} 5 & -2 \\ -9 & 4 \end{pmatrix}$

$$\text{The product} = \begin{pmatrix} (15-18) & (-6+8) \\ (-35+54) & (14-24) \end{pmatrix} = \begin{pmatrix} -3 & 2 \\ 19 & -10 \end{pmatrix}$$

Note that, from (2) and (3), the two products are not equal. Therefore, wher A and B are two matrices, in general, $A \times B$ is not necessarily equal to $B \times A$ *For matrices, multiplication is not commutative.*

Exercise 6.4

Calculate the following products:

1. $\begin{pmatrix} 1 & 2 \\ 3 & 4 \end{pmatrix} \cdot \begin{pmatrix} 1 & 3 \\ 2 & 4 \end{pmatrix}$

2. $\begin{pmatrix} 1 & 2 \\ 2 & 1 \end{pmatrix} \cdot \begin{pmatrix} 1 & 2 \\ 2 & 1 \end{pmatrix}$

3. $\begin{pmatrix} 2 & -1 \\ 1 & -2 \end{pmatrix} \cdot \begin{pmatrix} 2 & 1 \\ -1 & -2 \end{pmatrix}$

4. $\begin{pmatrix} 5 & 6 \\ 2 & -3 \end{pmatrix} \cdot \begin{pmatrix} 4 & 8 \\ 5 & 7 \end{pmatrix}$

5. $\begin{pmatrix} 4 & 8 \\ 5 & 7 \end{pmatrix} \cdot \begin{pmatrix} 5 & 6 \\ 2 & -3 \end{pmatrix}$

6. $\begin{pmatrix} 4 & -1 \\ -3 & 0 \end{pmatrix} \cdot \begin{pmatrix} -7 & 2 \\ 0 & -5 \end{pmatrix}$

. $\begin{pmatrix} -7 & 2 \\ 0 & -5 \end{pmatrix} \cdot \begin{pmatrix} 4 & -1 \\ -3 & 0 \end{pmatrix}$

. $\begin{pmatrix} a & b \\ b & a \end{pmatrix} \cdot \begin{pmatrix} c & d \\ d & c \end{pmatrix}$

. Calculate $(A \times B) \times C$ and $A \times (B \times C)$

where $A = \begin{pmatrix} 1 & 2 \\ 3 & 4 \end{pmatrix}$ $B = \begin{pmatrix} 1 & 3 \\ 2 & 4 \end{pmatrix}$ $C = \begin{pmatrix} 3 & 1 \\ 4 & 2 \end{pmatrix}$

0. $\begin{pmatrix} 15 & 6 \\ -12 & 4 \end{pmatrix} \cdot \begin{pmatrix} -13 & 13 \\ 17 & -9 \end{pmatrix}$

1. $\begin{pmatrix} -13 & 13 \\ 17 & -9 \end{pmatrix} \cdot \begin{pmatrix} 15 & 6 \\ -12 & 4 \end{pmatrix}$

2. $\begin{pmatrix} a & b \\ b & a \end{pmatrix} \cdot \begin{pmatrix} c & d \\ d & c \end{pmatrix}$

3. $\begin{pmatrix} c & d \\ d & c \end{pmatrix} \cdot \begin{pmatrix} a & b \\ b & a \end{pmatrix}$

4. $\begin{pmatrix} a & b \\ c & d \end{pmatrix} \cdot \begin{pmatrix} A & C \\ B & D \end{pmatrix}$

5. $\begin{pmatrix} A & C \\ B & D \end{pmatrix} \cdot \begin{pmatrix} a & b \\ c & d \end{pmatrix}$

6. $\begin{pmatrix} 1 & 0 \\ 0 & 1 \end{pmatrix} \cdot \begin{pmatrix} 3 & 5 \\ -2 & 7 \end{pmatrix}$

7. $\begin{pmatrix} 1 & 0 \\ 0 & 1 \end{pmatrix} \cdot \begin{pmatrix} -18 & -35 \\ 29 & -17 \end{pmatrix}$

8. $\begin{pmatrix} 3 & 5 \\ -2 & 7 \end{pmatrix} \cdot \begin{pmatrix} 1 & 0 \\ 0 & 1 \end{pmatrix}$

9. $\begin{pmatrix} -18 & -35 \\ 29 & -17 \end{pmatrix} \cdot \begin{pmatrix} 1 & 0 \\ 0 & 1 \end{pmatrix}$

0. $\begin{pmatrix} 1 & 0 \\ 0 & 1 \end{pmatrix} \cdot \begin{pmatrix} a & b \\ c & d \end{pmatrix}$

1. $\begin{pmatrix} a & b \\ c & d \end{pmatrix} \cdot \begin{pmatrix} 1 & 0 \\ 0 & 1 \end{pmatrix}$

ote the results for questions (4) and (5), (6) and (7), (10) and (11), (12) and

(13), (14) and (15). They endorse the fact that multiplication of matrices i not commutative. The results of questions (16), (17), (18), (19), (20) and (21 indicate that there seems to be something special about the matrix:

$$\begin{pmatrix} 1 & 0 \\ 0 & 1 \end{pmatrix}$$

Call it I. Questions (20) and (21) prove that when we multiply any othe matrix by it, in front or behind, then that matrix remains unchanged. I performs a function for matrices similar to that performed by 1 in th multiplication of real numbers. For that reason it is called the *unit matrix*, o the *identity matrix*. Where A represents any 2×2 matrix then:

$$I \times A = A \times I = A \qquad 6.$$

Or, in full:

$$\begin{pmatrix} 1 & 0 \\ 0 & 1 \end{pmatrix} \cdot \begin{pmatrix} a & b \\ c & d \end{pmatrix} = \begin{pmatrix} a & b \\ c & d \end{pmatrix} \cdot \begin{pmatrix} 1 & 0 \\ 0 & 1 \end{pmatrix} = \begin{pmatrix} a & b \\ c & d \end{pmatrix} \qquad 6.$$

Examples

Calculate the following:

1. $\begin{pmatrix} 2 & 3 \\ 1 & 2 \end{pmatrix} \cdot \begin{pmatrix} 2 & -3 \\ -1 & 2 \end{pmatrix}$

   The product $= \begin{pmatrix} (4-3) & (-6+6) \\ (2-2) & (-3+4) \end{pmatrix} = \begin{pmatrix} 1 & 0 \\ 0 & 1 \end{pmatrix}$

2. $\begin{pmatrix} -5 & 11 \\ 4 & -9 \end{pmatrix} \cdot \begin{pmatrix} -9 & -11 \\ -4 & -5 \end{pmatrix}$

   The product $= \begin{pmatrix} (45-44) & (55-55) \\ (-36+36) & (-44+45) \end{pmatrix} = \begin{pmatrix} 1 & 0 \\ 0 & 1 \end{pmatrix}$

3. $\begin{pmatrix} 2 & -3 \\ -1 & 2 \end{pmatrix} \cdot \begin{pmatrix} 2 & 3 \\ 1 & 2 \end{pmatrix}$

   The product $= \begin{pmatrix} (4-3) & (6-6) \\ (-2+2) & (-3+4) \end{pmatrix} = \begin{pmatrix} 1 & 0 \\ 0 & 1 \end{pmatrix}$

4. $\begin{pmatrix} -9 & -11 \\ -4 & -5 \end{pmatrix} \cdot \begin{pmatrix} -5 & 11 \\ 4 & -9 \end{pmatrix}$

$$\text{The product} = \begin{pmatrix} (45-44) & (-99+99) \\ (20-20) & (-44+45) \end{pmatrix} = \begin{pmatrix} 1 & 0 \\ 0 & 1 \end{pmatrix}$$

Note that, in each case, the product is the unit matrix.

Exercise 6.5

Calculate the following:

1. $\begin{pmatrix} 4 & 3 \\ 1 & 1 \end{pmatrix} \cdot \begin{pmatrix} 1 & -3 \\ -1 & 4 \end{pmatrix}$
2. $\begin{pmatrix} 1 & -3 \\ -1 & 4 \end{pmatrix} \cdot \begin{pmatrix} 4 & 3 \\ 1 & 1 \end{pmatrix}$
3. $\begin{pmatrix} 5 & 2 \\ 2 & 1 \end{pmatrix} \cdot \begin{pmatrix} 1 & -2 \\ -2 & 5 \end{pmatrix}$
4. $\begin{pmatrix} 1 & -2 \\ -2 & 5 \end{pmatrix} \cdot \begin{pmatrix} 5 & 2 \\ 2 & 1 \end{pmatrix}$
5. $\begin{pmatrix} 1 & 2 \\ 2 & 5 \end{pmatrix} \cdot \begin{pmatrix} 5 & -2 \\ -2 & 1 \end{pmatrix}$
6. $\begin{pmatrix} 5 & -2 \\ -2 & 1 \end{pmatrix} \cdot \begin{pmatrix} 1 & 2 \\ 2 & 5 \end{pmatrix}$
7. $\begin{pmatrix} 4 & 3 \\ 1 & 2 \end{pmatrix} \cdot \begin{pmatrix} 2 & -3 \\ -1 & 4 \end{pmatrix}$
8. $\begin{pmatrix} 2 & -3 \\ -1 & 4 \end{pmatrix} \cdot \begin{pmatrix} 4 & 3 \\ 1 & 2 \end{pmatrix}$
9. $\begin{pmatrix} 5 & 6 \\ 2 & 4 \end{pmatrix} \cdot \begin{pmatrix} 4 & -6 \\ -2 & 5 \end{pmatrix}$
10. $\begin{pmatrix} 4 & -6 \\ -2 & 5 \end{pmatrix} \cdot \begin{pmatrix} 5 & 6 \\ 2 & 4 \end{pmatrix}$
11. $\begin{pmatrix} -5 & -8 \\ -3 & -7 \end{pmatrix} \cdot \begin{pmatrix} -7 & 8 \\ 3 & -5 \end{pmatrix}$
12. $\begin{pmatrix} -7 & 8 \\ 3 & -5 \end{pmatrix} \cdot \begin{pmatrix} -5 & -8 \\ -3 & -7 \end{pmatrix}$

6.3 The determinant of a matrix

The results of the first six questions of exercise 6.5 give a product which is:

$$\begin{pmatrix} 1 & 0 \\ 0 & 1 \end{pmatrix}$$

which is the unit matrix, I.

The results of the last six questions give products which are a little different:

7 and 8 produce the answer $\begin{pmatrix} 5 & 0 \\ 0 & 5 \end{pmatrix}$

9 and 10 produce the answer $\begin{pmatrix} 8 & 0 \\ 0 & 8 \end{pmatrix}$

11 and 12 produce the answer $\begin{pmatrix} 11 & 0 \\ 0 & 11 \end{pmatrix}$

By formula *6.4*:

$$\begin{pmatrix} 5 & 0 \\ 0 & 5 \end{pmatrix} = 5\begin{pmatrix} 1 & 0 \\ 0 & 1 \end{pmatrix}$$

$$\begin{pmatrix} 8 & 0 \\ 0 & 8 \end{pmatrix} = 8\begin{pmatrix} 1 & 0 \\ 0 & 1 \end{pmatrix}$$

$$\begin{pmatrix} 11 & 0 \\ 0 & 11 \end{pmatrix} = 11\begin{pmatrix} 1 & 0 \\ 0 & 1 \end{pmatrix}$$

In other words, they give $5I$, $8I$ and $11I$ respectively. Therefore the products of certain matrices produce either I or a multiple of I. Where this property arises it is of great use in solving equations. This issue will be taken up in the next chapter. One aspect we need to clarify here is what determines whether the product is I or a multiple of I.

Associated with any square matrix is a numerical value which is important. For the matrix:

$$\begin{pmatrix} a & b \\ c & d \end{pmatrix}$$

that value is $ad - cb$, or the difference between the products of the pairs along the diagonals of the matrix, i.e.

$$\begin{matrix} a & & b \\ & \times & \\ c & & d \end{matrix}$$

This value is called the *determinant of the matrix*.

Examples

Calculate the determinants of the following matrices:

1. $\begin{pmatrix} 4 & 5 \\ 2 & 9 \end{pmatrix}$ The determinant $= 4 \times 9 - 2 \times 5 = 36 - 10 = 26$.

2. $\begin{pmatrix} 11 & 4 \\ -7 & 3 \end{pmatrix}$ The determinant $= 11 \times 3 - (-7 \times 4) = 33 + 28 = 61$.

3. $\begin{pmatrix} -3 & -2 \\ 8 & -5 \end{pmatrix}$ The determinant $= (-3 \times -5) - (8 \times -2) = 15 + 16$ $= 31$.

4. $\begin{pmatrix} 2 & 3 \\ 1 & 2 \end{pmatrix}$ The determinant $= 2 \times 2 - 1 \times 3 = 4 - 3 = 1$.

5. $\begin{pmatrix} -5 & 11 \\ 4 & -9 \end{pmatrix}$ The determinant $= (-5 \times -9) - 4 \times 11 = 45 - 44 = 1$.

6. $\begin{pmatrix} 4 & 3 \\ 1 & 2 \end{pmatrix}$ The determinant $= 4 \times 2 - 1 \times 3 = 8 - 3 = 5$.

Exercise 6.6

Calculate the determinants of the following matrices:

1. $\begin{pmatrix} 4 & 3 \\ 1 & 1 \end{pmatrix}$

2. $\begin{pmatrix} 1 & -3 \\ -1 & 4 \end{pmatrix}$

3. $\begin{pmatrix} 5 & 2 \\ 2 & 1 \end{pmatrix}$

4. $\begin{pmatrix} 4 & 3 \\ 1 & 2 \end{pmatrix}$

5. $\begin{pmatrix} 2 & -3 \\ -1 & 4 \end{pmatrix}$

6. $\begin{pmatrix} -9 & -11 \\ -4 & -5 \end{pmatrix}$

7. $\begin{pmatrix} 5 & 6 \\ 2 & 4 \end{pmatrix}$

8. $\begin{pmatrix} 4 & -6 \\ -2 & 5 \end{pmatrix}$

9. $\begin{pmatrix} 3 & 8 \\ 7 & 4 \end{pmatrix}$

10. $\begin{pmatrix} -8 & 4 \\ -2 & -3 \end{pmatrix}$

11. $\begin{pmatrix} a & -a \\ a & a \end{pmatrix}$

12. $\begin{pmatrix} 2a & 5a \\ a & 3a \end{pmatrix}$

13. $\begin{pmatrix} a & 1/b \\ -b & 1/a \end{pmatrix}$

14. $\begin{pmatrix} (a+b) & b \\ a & a \end{pmatrix}$

15. $\begin{pmatrix} a & 1 \\ -2a & 1 \end{pmatrix}$

In exercise 6.6 note those determinants whose values are 1. Relate them to the questions in exercise 6.5 which produced the unit matrix. Note further the occasions where the determinantal value coincided with the multiple of the unit matrix in exercise 6.5. The determinant itself has a special notation. The determinant of

$$\begin{pmatrix} a & b \\ c & d \end{pmatrix}$$

is variously written as:

$\det\begin{pmatrix} a & b \\ c & d \end{pmatrix}$, $\begin{vmatrix} a & b \\ c & d \end{vmatrix}$, and $ad-cb$, or $ad-bc$, and $|A|$, where A is the matrix

7

The Use of Matrices and Determinants in the Solution of Simultaneous Linear Equations in Two Unknowns

Example

Determine the solution of

$$\begin{aligned} 2x + y &= 5 \\ x + y &= 3 \end{aligned} \qquad \text{(i)}$$

In matrix notation (i) becomes:

$$\begin{pmatrix} 2 & 1 \\ 1 & 1 \end{pmatrix} \cdot \begin{pmatrix} x \\ y \end{pmatrix} = \begin{pmatrix} 5 \\ 3 \end{pmatrix} \qquad \text{(ii)}$$

Multiply (ii) by $\begin{pmatrix} 1 & -1 \\ -1 & 2 \end{pmatrix}$

$$\begin{pmatrix} 1 & -1 \\ -1 & 2 \end{pmatrix} \cdot \begin{pmatrix} 2 & 1 \\ 1 & 1 \end{pmatrix} \cdot \begin{pmatrix} x \\ y \end{pmatrix} = \begin{pmatrix} 1 & -1 \\ -1 & 2 \end{pmatrix} \cdot \begin{pmatrix} 5 \\ 3 \end{pmatrix} \qquad \text{(iii)}$$

$$\begin{pmatrix} 1 & 0 \\ 0 & 1 \end{pmatrix} \cdot \begin{pmatrix} x \\ y \end{pmatrix} = \begin{pmatrix} 2 \\ 1 \end{pmatrix} \qquad \text{(iv)}$$

$$\begin{pmatrix} x \\ y \end{pmatrix} = \begin{pmatrix} 2 \\ 1 \end{pmatrix} \qquad \text{(v)}$$

(v) means that $x = 2$, $y = 1$.
Check: substitute back in (i):
LHS of the first equation $= 2 \times 2 + 1 = 5$
LHS of the second equation $= 2 + 1 = 3$

A second check is to solve the equations by either elimination or substitution. The method above solves the equations in (i).

The important step in that solution is the use of the matrix

$$\begin{pmatrix} 1 & -1 \\ -1 & 2 \end{pmatrix}$$

for the multiplication of (ii). We need to know how to determine that matrix which produces the solution of a set of equations when, in fact, the equations are soluble. Suppose we write the matrix

$$\begin{pmatrix} 2 & 1 \\ 1 & 1 \end{pmatrix}$$

as A. Then we write the matrix

$$\begin{pmatrix} 1 & -1 \\ -1 & 2 \end{pmatrix}$$

as A^{-1}, because $A^{-1}.A = I$, the unit matrix. Associate this step in matrices with the step in real numbers:

$$\frac{1}{a} \times a = 1, \text{ or } a^{-1}.a = 1$$

A^{-1} is called the inverse of the matrix A. It is also true that $A.A^{-1} = I$:

$$\begin{pmatrix} 2 & 1 \\ 1 & 1 \end{pmatrix} . \begin{pmatrix} 1 & -1 \\ -1 & 2 \end{pmatrix} = \begin{pmatrix} 1 & 0 \\ 0 & 1 \end{pmatrix}$$

Each matrix is the inverse of the other: A^{-1} is the inverse of A, and A is the inverse of A^{-1}.

7.1 The inverse of a 2 × 2 matrix

Definition

Matrices which are inverses of each other are such that their product, in either order, is the unit matrix.

Examples

Show that the following pairs of matrices are inverses of each other. Note carefully the relationship of the elements of one to the elements of the other.

1. $\begin{pmatrix} 3 & 5 \\ 1 & 2 \end{pmatrix}, \begin{pmatrix} 2 & -5 \\ -1 & 3 \end{pmatrix}$

   The product $\begin{pmatrix} 3 & 5 \\ 1 & 2 \end{pmatrix} \cdot \begin{pmatrix} 2 & -5 \\ -1 & 3 \end{pmatrix} = \begin{pmatrix} 1 & 0 \\ 0 & 1 \end{pmatrix}$

   The product $\begin{pmatrix} 2 & -5 \\ -1 & 3 \end{pmatrix} \cdot \begin{pmatrix} 3 & 5 \\ 1 & 2 \end{pmatrix} = \begin{pmatrix} 1 & 0 \\ 0 & 1 \end{pmatrix}$

2. $\begin{pmatrix} 8 & 5 \\ 3 & 2 \end{pmatrix}, \begin{pmatrix} 2 & -5 \\ -3 & 8 \end{pmatrix}$

   The product $\begin{pmatrix} 8 & 5 \\ 3 & 2 \end{pmatrix} \cdot \begin{pmatrix} 2 & -5 \\ -3 & 8 \end{pmatrix} = \begin{pmatrix} 1 & 0 \\ 0 & 1 \end{pmatrix}$

   The product $\begin{pmatrix} 2 & -5 \\ -3 & 8 \end{pmatrix} \cdot \begin{pmatrix} 8 & 5 \\ 3 & 2 \end{pmatrix} = \begin{pmatrix} 1 & 0 \\ 0 & 1 \end{pmatrix}$

Exercise 7.1

Show that the following pairs of matrices are inverses of each other. Note carefully the relationship of the elements of one to the corresponding elements of the other.

1. $\begin{pmatrix} 3 & 1 \\ 2 & 1 \end{pmatrix}, \begin{pmatrix} 1 & -1 \\ -2 & 3 \end{pmatrix}$
2. $\begin{pmatrix} 1 & 1 \\ 2 & 3 \end{pmatrix}, \begin{pmatrix} 3 & -1 \\ -2 & 1 \end{pmatrix}$
3. $\begin{pmatrix} 5 & 2 \\ 2 & 1 \end{pmatrix}, \begin{pmatrix} 1 & -2 \\ -2 & 5 \end{pmatrix}$
4. $\begin{pmatrix} 1 & 1 \\ 4 & 5 \end{pmatrix}, \begin{pmatrix} 5 & -1 \\ -4 & 1 \end{pmatrix}$
5. $\begin{pmatrix} 4 & -7 \\ -1 & 2 \end{pmatrix}, \begin{pmatrix} 2 & 7 \\ 1 & 4 \end{pmatrix}$
6. $\begin{pmatrix} -9 & -7 \\ -5 & -4 \end{pmatrix}, \begin{pmatrix} -4 & 7 \\ 5 & -9 \end{pmatrix}$

7. $\begin{pmatrix} 27 & 5 \\ 16 & 3 \end{pmatrix}, \begin{pmatrix} 3 & -5 \\ -16 & 27 \end{pmatrix}$

8. $\begin{pmatrix} -8 & -9 \\ 9 & 10 \end{pmatrix}, \begin{pmatrix} 10 & 9 \\ -9 & -8 \end{pmatrix}$

By observation of the above examples it appears that the inverse of

$$\begin{pmatrix} a & b \\ c & d \end{pmatrix}$$

might be

$$\begin{pmatrix} d & -b \\ -c & a \end{pmatrix}$$

Examples

1. Check whether the matrices

$$\begin{pmatrix} 3 & 1 \\ 1 & 2 \end{pmatrix}, \begin{pmatrix} 2 & -1 \\ -1 & 3 \end{pmatrix}$$

are inverses of each other.

The product $\begin{pmatrix} 3 & 1 \\ 1 & 2 \end{pmatrix} \cdot \begin{pmatrix} 2 & -1 \\ -1 & 3 \end{pmatrix} = \begin{pmatrix} 5 & 0 \\ 0 & 5 \end{pmatrix}$

The product $\begin{pmatrix} 2 & -1 \\ -1 & 3 \end{pmatrix} \cdot \begin{pmatrix} 3 & 1 \\ 1 & 2 \end{pmatrix} = \begin{pmatrix} 5 & 0 \\ 0 & 5 \end{pmatrix}$

The products are not the unit matrix. However, they are closely related to I:

$$\begin{pmatrix} 5 & 0 \\ 0 & 5 \end{pmatrix} = 5 \begin{pmatrix} 1 & 0 \\ 0 & 1 \end{pmatrix}$$

Moreover $5 = \det \begin{pmatrix} 3 & 1 \\ 1 & 2 \end{pmatrix}$ i.e. $\begin{vmatrix} 3 & 1 \\ 1 & 2 \end{vmatrix} = 3 \times 2 - 1^2$

2. In view of the result of (1) check whether

$$\frac{1}{5} \begin{pmatrix} 2 & -1 \\ -1 & 3 \end{pmatrix}$$

is the inverse of

$$\begin{pmatrix} 3 & 1 \\ 1 & 2 \end{pmatrix}$$

Step 1. $\frac{1}{5}\begin{pmatrix} 2 & -1 \\ -1 & 3 \end{pmatrix} = \begin{pmatrix} 2/5 & -1/5 \\ -1/5 & 3/5 \end{pmatrix}$

Step 2. $\begin{pmatrix} 3 & 1 \\ 1 & 2 \end{pmatrix} \cdot \begin{pmatrix} 2/5 & -1/5 \\ -1/5 & 3/5 \end{pmatrix} = \begin{pmatrix} 1 & 0 \\ 0 & 1 \end{pmatrix}$

Step 3. $\begin{pmatrix} 2/5 & -1/5 \\ -1/5 & 3/5 \end{pmatrix} \cdot \begin{pmatrix} 3 & 1 \\ 1 & 2 \end{pmatrix} = \begin{pmatrix} 1 & 0 \\ 0 & 1 \end{pmatrix}$

It is the inverse.

Exercise 7.2

Show that the following pairs of matrices are inverses of each other. Check that the denominator of each element in the second matrix equals the determinant of the first matrix.

1. $\begin{pmatrix} 3 & 1 \\ 1 & 1 \end{pmatrix}\begin{pmatrix} 1/2 & -1/2 \\ -1/2 & 3/2 \end{pmatrix}$

2. $\begin{pmatrix} 4 & 1 \\ 1 & 1 \end{pmatrix}\begin{pmatrix} 1/3 & -1/3 \\ -1/3 & 4/3 \end{pmatrix}$

3. $\begin{pmatrix} 5 & 2 \\ 1 & 1 \end{pmatrix}\begin{pmatrix} 1/3 & -2/3 \\ -1/3 & 5/3 \end{pmatrix}$

4. $\begin{pmatrix} 11 & 3 \\ 5 & 2 \end{pmatrix}\begin{pmatrix} 2/7 & -3/7 \\ -5/7 & 11/7 \end{pmatrix}$

5. $\begin{pmatrix} 3 & 1 \\ -1 & 1 \end{pmatrix}\begin{pmatrix} 1/4 & -1/4 \\ 1/4 & 3/4 \end{pmatrix}$

6. $\begin{pmatrix} 7 & 3 \\ -2 & 5 \end{pmatrix}\begin{pmatrix} 5/41 & -3/41 \\ 2/41 & 7/41 \end{pmatrix}$

7. $\begin{pmatrix} a & 0 \\ 1 & 1 \end{pmatrix}\begin{pmatrix} 1/a & 0 \\ -1/a & 1 \end{pmatrix}$

8. $\begin{pmatrix} a & 0 \\ 1 & b \end{pmatrix}\begin{pmatrix} b/ab & 0 \\ -1/ab & a/ab \end{pmatrix}$

9. $\begin{pmatrix} a & 1 \\ c & b \end{pmatrix}\begin{pmatrix} b/(ab-c) & -1/(ab-c) \\ -c/(ab-c) & a/(ab-c) \end{pmatrix}$

10. $\begin{pmatrix} a & b \\ c & d \end{pmatrix} \begin{pmatrix} \dfrac{d}{ad-bc} & \dfrac{-b}{ad-bc} \\ \dfrac{-c}{ad-bc} & \dfrac{a}{ad-bc} \end{pmatrix}$

Question (10) in exercise 7.2 provides a rule for writing down the inverse of any 2×2 matrix when an inverse exists.

Definition

The inverse of the matrix $A = \begin{pmatrix} a & b \\ c & d \end{pmatrix}$ is the matrix $\begin{pmatrix} d/\Delta & -b/\Delta \\ -c/\Delta & a/\Delta \end{pmatrix}$ where $\Delta = ad - bc$, the determinant of A.

Examples

1. Determine the inverse of $A = \begin{pmatrix} 7 & 3 \\ 4 & 5 \end{pmatrix}$

Step 1. Det $A = 7 \times 5 - 4 \times 3 = 35 - 12 = 23$.

Step 2. $A^{-1} = \begin{pmatrix} 5/23 & -3/23 \\ -4/23 & 7/23 \end{pmatrix}$

Check: $\begin{pmatrix} 7 & 3 \\ 4 & 5 \end{pmatrix} \cdot \begin{pmatrix} 5/23 & -3/23 \\ -4/23 & 7/23 \end{pmatrix} = \begin{pmatrix} 23/23 & 0 \\ 0 & 23/23 \end{pmatrix} = \begin{pmatrix} 1 & 0 \\ 0 & 1 \end{pmatrix}$

2. Determine the inverse of $B = \begin{pmatrix} 11 & 5 \\ -18 & 17 \end{pmatrix}$

Step 1. Det $B = 11 \times 17 - (-18 \times 5) = 187 + 90 = 277$.

Step 2. $B^{-1} = \begin{pmatrix} 17/277 & -5/277 \\ 18/277 & 11/277 \end{pmatrix}$

Check: $\begin{pmatrix} 11 & 5 \\ -18 & 17 \end{pmatrix} \cdot \begin{pmatrix} 17/277 & -5/277 \\ 18/277 & 11/277 \end{pmatrix} = \begin{pmatrix} 277/277 & 0 \\ 0 & 277/277 \end{pmatrix}$

$$= \begin{pmatrix} 1 & 0 \\ 0 & 1 \end{pmatrix}$$

Exercise 7.3

Determine the inverses of the following matrices:

1. $\begin{pmatrix} 3 & 2 \\ 5 & 4 \end{pmatrix}$

2. $\begin{pmatrix} 4 & 1 \\ 3 & 2 \end{pmatrix}$

3. $\begin{pmatrix} 7 & 3 \\ 3 & 2 \end{pmatrix}$

4. $\begin{pmatrix} 9 & 3 \\ 5 & 4 \end{pmatrix}$

5. $\begin{pmatrix} 4 & 5 \\ -2 & 3 \end{pmatrix}$

6. $\begin{pmatrix} -5 & 3 \\ -4 & 2 \end{pmatrix}$

7. $\begin{pmatrix} -1 & 2 \\ -4 & 3 \end{pmatrix}$

8. $\begin{pmatrix} -3 & 2 \\ -7 & -3 \end{pmatrix}$

9. $\begin{pmatrix} a & b \\ -a & b \end{pmatrix}$

10. $\begin{pmatrix} 7 & 12 \\ 11 & 5 \end{pmatrix}$

7.2 The solution of simultaneous linear equations by matrices

Examples

Solve the following pairs of equations:

1.
$$\begin{aligned} 5x + 9y &= 17 \\ 3x + 8y &= 11 \end{aligned} \qquad \text{(i)}$$

Step 1. Rewrite (i):

$$\begin{pmatrix} 5 & 9 \\ 3 & 8 \end{pmatrix} \cdot \begin{pmatrix} x \\ y \end{pmatrix} = \begin{pmatrix} 17 \\ 11 \end{pmatrix} \tag{ii}$$

Step 2. $\text{Det}\begin{pmatrix} 5 & 9 \\ 3 & 8 \end{pmatrix} = 40 - 27 = 13.$

Step 3. The inverse of $\begin{pmatrix} 5 & 9 \\ 3 & 8 \end{pmatrix} = \begin{pmatrix} 8/13 & -9/13 \\ -3/13 & 5/13 \end{pmatrix}$

Step 4. Multiply (ii) by this inverse:

$$\begin{pmatrix} 8/13 & -9/13 \\ -3/13 & 5/13 \end{pmatrix} \cdot \begin{pmatrix} 5 & 9 \\ 3 & 8 \end{pmatrix} \cdot \begin{pmatrix} x \\ y \end{pmatrix} = \begin{pmatrix} 8/13 & -9/13 \\ -3/13 & 5/13 \end{pmatrix} \cdot \begin{pmatrix} 17 \\ 11 \end{pmatrix}$$

$$\begin{pmatrix} 1 & 0 \\ 0 & 1 \end{pmatrix} \cdot \begin{pmatrix} x \\ y \end{pmatrix} = \begin{pmatrix} 37/13 \\ 4/13 \end{pmatrix}$$

Step 5. $\begin{pmatrix} x \\ y \end{pmatrix} = \begin{pmatrix} 37/13 \\ 4/13 \end{pmatrix}$

Step 6. $x = 37/13$; $y = 4/13$.
Check: Substitute in (i):

$$5 \times \frac{37}{13} + 9 \times \frac{4}{13} = \frac{185}{13} + \frac{36}{13} = 221/13 = 17$$

$$3 \times \frac{37}{13} + 8 \times \frac{4}{13} = \frac{111}{13} + \frac{32}{13} = \frac{143}{13} = 11$$

2.

$$\begin{aligned} 23x - 12y &= 37 \\ 4x + \ 5y &= -5 \end{aligned} \tag{i}$$

Step 1. Rewrite (i):

$$\begin{pmatrix} 23 & -12 \\ 4 & 5 \end{pmatrix} \cdot \begin{pmatrix} x \\ y \end{pmatrix} = \begin{pmatrix} 37 \\ -5 \end{pmatrix} \tag{ii}$$

Step 2. $\text{Det}\begin{pmatrix} 23 & -12 \\ 4 & 5 \end{pmatrix} = 115 + 48 = 163$

Step 3. The inverse of $\begin{pmatrix} 23 & -12 \\ 4 & 5 \end{pmatrix} = \begin{pmatrix} 5/163 & 12/163 \\ -4/163 & 23/163 \end{pmatrix}$

Step 4. Multiply (ii) by this inverse:

$$\begin{pmatrix} 5/163 & 12/163 \\ -4/163 & 23/163 \end{pmatrix} \cdot \begin{pmatrix} 23 & -12 \\ 4 & 5 \end{pmatrix} \cdot \begin{pmatrix} x \\ y \end{pmatrix} = \begin{pmatrix} 5/163 & 12/163 \\ -4/163 & 23/163 \end{pmatrix} \cdot \begin{pmatrix} 37 \\ -5 \end{pmatrix}$$

$$\begin{pmatrix} 1 & 0 \\ 0 & 1 \end{pmatrix} \cdot \begin{pmatrix} x \\ y \end{pmatrix} = \begin{pmatrix} 125/163 \\ -263/163 \end{pmatrix}$$

Step 5. $\begin{pmatrix} x \\ y \end{pmatrix} = \begin{pmatrix} 125/163 \\ -263/163 \end{pmatrix}$

Step 6. $x = 125/163$; $y = -263/163$.
Check: Substitute in (i):

$$23 \times \frac{125}{163} - 12\left(-\frac{263}{163}\right) = \frac{2875}{163} + \frac{3156}{163} = \frac{6031}{163} = 37$$

$$4 \times \frac{125}{163} + 5\left(-\frac{263}{163}\right) = \frac{500}{163} - \frac{1315}{163} = -\frac{815}{163} = -5$$

3.
$$\begin{aligned} ax + by &= p \\ cx + dy &= q \end{aligned} \qquad \text{(i)}$$

Step 1. Rewrite (i):

$$\begin{pmatrix} a & b \\ c & d \end{pmatrix} \cdot \begin{pmatrix} x \\ y \end{pmatrix} = \begin{pmatrix} p \\ q \end{pmatrix} \qquad \text{(ii)}$$

Step 2. $\text{Det}\begin{pmatrix} a & b \\ c & d \end{pmatrix} = ad - bc = \Delta$

Step 3. The inverse of $\begin{pmatrix} a & b \\ c & d \end{pmatrix} = \begin{pmatrix} d/\Delta & -b/\Delta \\ -c/\Delta & a/\Delta \end{pmatrix}$

Step 4. Multiply (ii) by this inverse:

$$\begin{pmatrix} d/\Delta & -b/\Delta \\ -c/\Delta & a/\Delta \end{pmatrix} \cdot \begin{pmatrix} a & b \\ c & d \end{pmatrix} \cdot \begin{pmatrix} x \\ y \end{pmatrix} = \begin{pmatrix} d/\Delta & -b/\Delta \\ -c/\Delta & a/\Delta \end{pmatrix} \cdot \begin{pmatrix} p \\ q \end{pmatrix}$$

$$\begin{pmatrix} 1 & 0 \\ 0 & 1 \end{pmatrix} \cdot \begin{pmatrix} x \\ y \end{pmatrix} = \begin{pmatrix} (pd - bq)/\Delta \\ (aq - cp)/\Delta \end{pmatrix}$$

Step 5. $\begin{pmatrix} x \\ y \end{pmatrix} = \begin{pmatrix} (dp - bq)/\Delta \\ (aq - cp)/\Delta \end{pmatrix}$

Step 6. $x = (dp - bq)/\Delta$; $y = (aq - cp)/\Delta$ *7.1*

Exercise 7.4

Solve the following pairs of equations by the matrix method:

1. $x + 2y = 3$
   $2x + y = 12$
2. $5x - 2y = 13$
   $3x + 2y = 11$

3. $3x-2y = -11$
   $3x-7y = -14$
4. $3a+2b = 11$
   $2a-b = 5$
5. $9x+5y = 17$
   $3x+2y = -1$
6. $6p-9q = -7$
   $9p-7q = -8$
7. $3x-4y = 0$
   $2x+4y = 13$

7.3 The solution of simultaneous linear equations by determinants

An alternative to the matrix method of solution of simultaneous linear equations is the determinant method. Example (3) in section 7.2 provides the basis for the alternative solution. From (i) the arrangement of the coefficients only in the two equations may be written in the form of an array, i.e. a 2×3 matrix, as follows:

$$\begin{pmatrix} a & b & p \\ c & d & q \end{pmatrix}$$

From this array we may extract three 2×2 matrices by excluding, in turn, one of the columns. These three matrices are:

$$\begin{pmatrix} a & b \\ c & d \end{pmatrix} \quad \text{omitting column 3}$$

$$\begin{pmatrix} b & p \\ d & q \end{pmatrix} \quad \text{omitting column 1}$$

$$\begin{pmatrix} a & p \\ c & q \end{pmatrix} \quad \text{omitting column 2}$$

Their determinants are, respectively, $(ad-cb) = \Delta$; $(bq-dp)$, call it Δ_1 $(aq-cp)$, call it Δ_2. Using this final notation the solution to Example (3 may be expressed:

$$x = -\Delta_1/\Delta; \; y = \Delta_2/\Delta \qquad 7.2$$

The following examples illustrate the method. As a check the data of examples (1) and (2) in section 7.2 will be used again.

Examples

1. Solve
$$5p+9q = 17$$
$$3p+8q = 11$$

Step 1. Write down the array:

$$\begin{pmatrix} 5 & 9 & 17 \\ 3 & 8 & 11 \end{pmatrix}$$

Step 2.

$$\Delta = \begin{vmatrix} 5 & 9 \\ 3 & 8 \end{vmatrix} = 40-27 = 13$$

$$\Delta_1 = \begin{vmatrix} 9 & 17 \\ 8 & 11 \end{vmatrix} = 99-136 = -37$$

$$\Delta_2 = \begin{vmatrix} 5 & 17 \\ 3 & 11 \end{vmatrix} = 55-51 = 4$$

Step 3. Substitute in $x = -\Delta_1/\Delta$; $y = \Delta_2/\Delta$:

$$x = 37/13; \quad y = 4/13$$

2. Solve
$$23a-12b = 37$$
$$4a+5b = -5$$

Step 1. Write down the array:

$$\begin{pmatrix} 23 & -12 & 37 \\ 4 & 5 & -5 \end{pmatrix}$$

Step 2.

$$\Delta = \begin{vmatrix} 23 & -12 \\ 4 & 5 \end{vmatrix} = 115+48 = 163$$

$$\Delta_1 = \begin{vmatrix} -12 & 37 \\ 5 & -5 \end{vmatrix} = 60-185 = -125$$

$$\Delta_2 = \begin{vmatrix} 23 & 37 \\ 4 & -5 \end{vmatrix} = -115-148 = -263$$

Step 3. Substitute in $x = -\Delta_1/\Delta$; $y = \Delta_2/\Delta$:

$$x = 125/163; \quad y = -263/163$$

Exercise 7.5

Solve the following pairs of equations by the determinant method. Check the answers to questions (1) to (7) by those obtained in Exercise 7.4.

1. $p+2q=3$
   $2p+q=12$
2. $5a-2b=13$
   $3a+2b=11$
3. $3p-5q=-11$
   $3p-7q=-14$
4. $3x+2y=11$
   $2x-y=5$
5. $9x+5y=17$
   $3x+2y=-1$
6. $6a-9b=-7$
   $9a-7b=-8$
7. $3x-4y=0$
   $2x+4y=13$
8. $7x-4y=44$
   $6x+3y=57$
9. $4x-6y=3/2$
   $7x-5y=27/4$
10. $14x+3y=7$
    $28x-2y=6$
11. $\frac{2}{3}x+\frac{5}{9}y=11$
    $6x+5y=46$
12. $\frac{1}{5}a-\frac{3}{10}b=10$
    $6a-9b=20$

Now try to solve the following equations by both the matrix method and the determinant method:

13. $4x+6y=19$
    $2x+3y=17$
14. $4x+6y=30$
    $2x+3y=15$
15. $ax+by=c$
    $pax+pby=d$
16. $ax+by=c$
    $pax+pby=pc$

Both the matrix method and the determinant method fail when we try to apply them to questions (11) to (16) in the last exercise. The determinant of the matrix on the left of the set of equations in each case is 0. Consequently the inverse does not exist; each element is indeterminate. For example, when we try to write down the inverse of

$$\begin{pmatrix} 4 & 6 \\ 2 & 3 \end{pmatrix} \text{ we produce } \begin{pmatrix} 3/0 & -6/0 \\ -2/0 & 4/0 \end{pmatrix}$$

Division by 0 is meaningless: none of the four elements can be determined.

7.4 A singular matrix

Definition

Any matrix whose determinant is zero is called a *singular matrix.*

Case 1

In both question 14 and question 16 in Exercise 7.5 the two equations are really identical, e.g. $4x + 6y = 30$ and $2x + 3y = 15$ both represent the same straight line:

$$y = -\tfrac{2}{3}x + 5$$

The number of solutions must be infinitely large.

Case 2

In both question 13 and question 15 in Exercise 7.5 the two equations represent lines which are parallel, e.g. $pax + pby = d$ is $ax + by = d/p$, which is a line parallel to:

$$ax + by = c$$

Two such lines will never meet. An alternative view is that they do meet but at an infinitely great distance away. Either view leads to the conclusion that there is no solution to the equations.

Examples

Show that the following matrices are singular:

1. $\begin{pmatrix} 5 & 1 \\ 10 & 2 \end{pmatrix}$ The determinant $= 5 \times 2 - 10 \times 1 = 10 - 10 = 0$

2. $\begin{pmatrix} -6 & -2 \\ 9 & 3 \end{pmatrix}$ The determinant $= (-6)(3) - (-2)(9) = -18 + 18 = 0$

3. $\begin{pmatrix} a & a^2 \\ b & ab \end{pmatrix}$ The determinant $= a(ab) - b(a^2) = a^2b - a^2b = 0$

4. Determine the condition for the following matrix to be singular:

$$\begin{pmatrix} a & b \\ b & a \end{pmatrix}$$

The determinant $= a^2 - b^2 = (a-b)(a+b)$. When $a - b = 0$, or $a + b = 0$, the determinant $= 0$, i.e. either $a = b$, or $a = -b$ for the matrix to be singular.

5. What values of a will make the following matrix singular?

$$\begin{pmatrix} a & 1 \\ 1 & a^2 \end{pmatrix}$$

The determinant $= a^3 - 1 = (a-1)(a^2 + a + 1)$. When $a = 1$ the determinant $= 0$. No other real value of a will make the determinant vanish. The matrix is singular only when $a = 1$.

6. Determine the conditions for the solution of the following pair of equations to be unique:

$$ax + by = c$$
$$bx + ay = d$$

For the solution to be unique:

$$\begin{vmatrix} a & b \\ b & a \end{vmatrix} \neq 0$$

$$a^2 - b^2 \neq 0$$

$$a \neq \pm b$$

Exercise 7.6

Determine which of the following matrices are singular:

1. $\begin{pmatrix} 4 & 2 \\ 2 & 1 \end{pmatrix}$

2. $\begin{pmatrix} -2 & 4 \\ 1 & 2 \end{pmatrix}$

3. $\begin{pmatrix} 6 & 3 \\ 8 & 4 \end{pmatrix}$

4. $\begin{pmatrix} 6 & 3 \\ -8 & -4 \end{pmatrix}$

5. $\begin{pmatrix} 7 & -9 \\ -21 & 27 \end{pmatrix}$

6. $\begin{pmatrix} 1/2 & 1/3 \\ 1/3 & 1/2 \end{pmatrix}$

7. $\begin{pmatrix} 2a & 2b \\ -3a & -3b \end{pmatrix}$

8. $\begin{pmatrix} a/b & 1 \\ -1 & d/a \end{pmatrix}$

Determine the conditions for the following matrices to be singular:

9. $\begin{pmatrix} p & 1 \\ 1 & p \end{pmatrix}$

10. $\begin{pmatrix} q & 3 \\ 5 & 4 \end{pmatrix}$

11. $\begin{pmatrix} 7 & -p \\ 4 & 5 \end{pmatrix}$

12. $\begin{pmatrix} p & 3 \\ 27 & p \end{pmatrix}$

13. $\begin{pmatrix} a^2 & 5 \\ 25 & a \end{pmatrix}$

14. $\begin{pmatrix} a & 5 \\ -25 & a^2 \end{pmatrix}$

15. $\begin{pmatrix} 2x & 3y \\ 4 & -2 \end{pmatrix}$

Determine the conditions in which the following pairs of equations have a unique solution:

16. $px + 3y = 7$
    $4x - 2y = 9$

17. $3x - ay = 12$
    $5x + 2y = 17$

18. $ax - by = c$
    $bx + ay = d$

Exercise 7.7

1. Currents i_1 and i_2 in a network are related by the equations:

$$i_1 + 2i_2 = 4.8$$
$$3i_1 - i_2 = 4.3$$

   Determine the values of i_1 and i_2.

2. The potential differences V_1 and V_2 along two parts of a circuit are given by:

$$1.2\,V_1 + 0.5\,V_2 = 72.2$$
$$0.8\,V_1 + 1.1\,V_2 = 74.2$$

   Determine V_1 and V_2.

3. Solve the following equations for i_1 and i_2:

$$0.1\,i_1 + 2(i_1 + i_2) = 4.1$$
$$0.04\,i_2 + (i_1 + i_2) = 2.15$$

4. The law relating effort, P, to load, W, for a particular machine is of the form $P = aW + b$. Given that when W is 1000 N P is 150 N and that when W is 10 000 N P is 750 N, determine the values of a and b.
5. The wind resistance, R, on a particular vehicle is related to the velocity, v, of the vehicle by the formula $R = p + qv^3$. Determine the values of p and q given that R is 5000 N when v is 20 m/s and R is 10 000 N when v is 30 m/s.
6. Network voltages V_1 and V_2 are given by:

$$3\,V_1 = 4\,V_2 + 185$$
$$5\,V_1 + 3\,V_2 = 647$$

   Determine the values of V_1 and V_2.

7. The resistances r_1 and r_2 in a network satisfy the relations:

$$1.4r_1 + 1.3r_2 = 14.8$$
$$3r_1 - 2.7r_2 = 3.8$$

   Determine r_1 and r_2.

8

The Rate of Change of a Function

8.1 The gradients of lines

The calculation of gradients of curves is based entirely on the gradients of lines. It is essential for the proper understanding of later work not only that the calculation of the gradients of straight lines is fully appreciated but also that what a gradient means is fully understood.

In Fig. 8.1 AB represents a straight line. A straight line has the same direction at every point of its length. This means it has a constant gradient at every point of its length. In earlier sections two methods of calculating gradients were discussed.

Method 1

When the equation of the line is known, e.g. $y = \frac{1}{2}x - 1$, then the gradient is $\frac{1}{2}$. In general a line whose equation is $y = mx + c$ has a gradient m (the coefficient of x in the equation when it is written in the standard form).

Method 2

When two points and their co-ordinates are known, say $(0, -1)$ and $(2, 0)$, then the gradient of the line joining them is:

$$\frac{0-(-1)}{2-0} = \frac{0+1}{2-0} = \frac{1}{2}$$

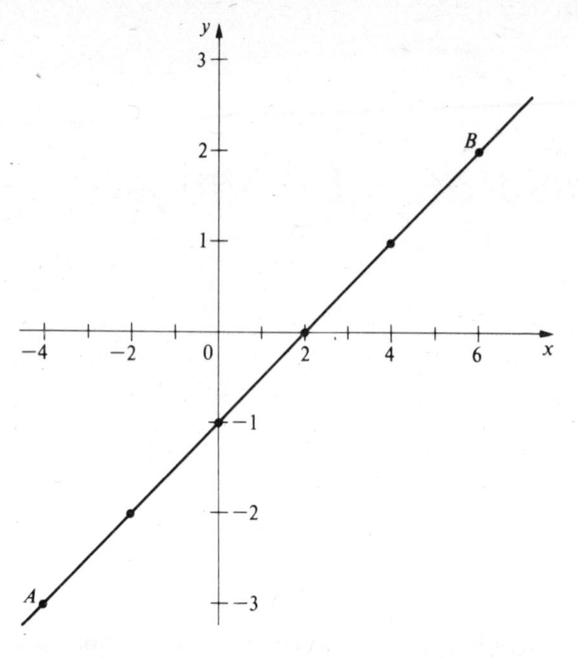

Figure 8.1

The co-ordinates of any two points on the line may be used. For example, from Fig. 8.1, the points (4, 1), (6, 2), (−2, −2) and (−4, −3) also lie on the line. When the points (−4, −3) and (4, 1) are taken as a pair the gradient of the line is:

$$\frac{1-(-3)}{4-(-4)} = \frac{1+3}{4+4} = \frac{4}{8} = \frac{1}{2}$$

By choosing other pairs of points from the selection above, the gradient, calculated in the above manner, always equals $\frac{1}{2}$. The formula for the gradient of a line which passes through (x_1, y_1), (x_2, y_2) is:

$$\frac{y_2 - y_1}{x_2 - x_1} \qquad 8.1$$

Fig. 8.2 is a slightly different approach from that of Fig. 8.1. By tabulating the details calculated above we obtain:

Pairs of points chosen	*Value of* $(x_2 - x_1)$	*Value of* $(y_2 - y_1)$
P, P_1	2	1
P, P_2	4	2

Pairs of points chosen	*Value of* $(x_2 - x_1)$	*Value of* $(y_2 - y_1)$
P, P_3	6	3
P, P_4	8	4
P, P_5	10	5

The number in the third column is always half that in the second column.

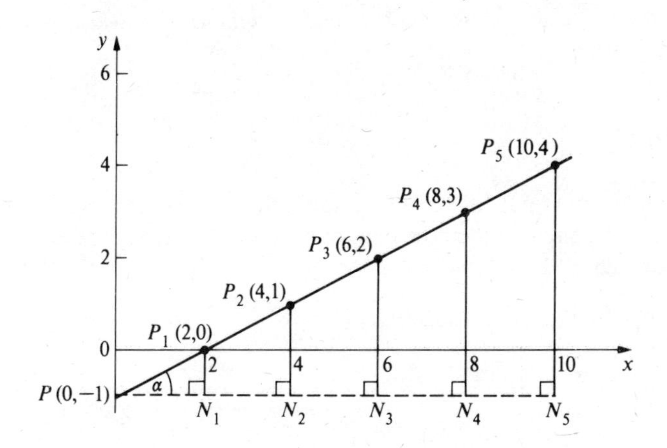

Figure 8.2

This can easily be related to trigonometry. Because all the triangles $PP_1N_1, PP_2N_2, PP_3N_3$, etc. have a common angle, α, $\tan \alpha = P_1N_1/PN_1 = P_2N_2/PN_2 = P_3N_3/PN_3 = P_4N_4/PN_4 = P_5N_5/PN_5$.

The gradient of the line is the tangent of the angle which the line makes with the positive direction of the x axis, Ox.

Examples

1. The gradient of the line joining $(3, 4)$ and $(8, 7)$ is

$$\frac{7-4}{8-3} = \frac{3}{5} = 0.6$$

If α is the angle which the line makes with Ox, then $\tan \alpha = 0.6000$; $\alpha = 30.963757 \approx 30°\ 58'$.

2. The gradient of the line joining $(-2, 3)$ and $(3, -4)$ is

$$\frac{-4-3}{-3-(-2)} = \frac{-7}{3+2} = \frac{-7}{5} = -1.4$$

If β is the angle which the line makes with Ox, then $\tan \beta = -1.4000$; $\beta = 180° - (54.462322°) = 125.53768° \approx 125°\ 32'$.

3. The gradient of the line joining $(-1, -4)$ and $(-9, -11)$ is

$$\frac{-11-(-4)}{-9-(-1)} = \frac{-11+4}{-9+1} = \frac{-7}{-8} = 7/8 = 0.875$$

If γ is the angle which the line makes with Ox, then $\tan \gamma = 0.8750$; $\gamma = 41.185925° \approx 41°\ 11'$.

Exercise 8.1

Calculate the gradients of the lines joining the following pairs of points and determine the angles which the lines make with Ox. Check the answers by a rough sketch.

1. $(0, 0)$, $(2, 3)$
2. $(0, 0)$, $(-2, 1)$
3. $(0, 0)$, $(-5, -2)$
4. $(0, 0)$, $(4, -3)$
5. $(2, 1)$, $(4, -3)$
6. $(5, 6)$, $(2, 1)$
7. $(4, 3)$, $(-5, 2)$
8. $(6, -4)$, $(-7, 2)$
9. $(-3, -2)$, $(-5, -6)$

Calculate the gradients of the lines joining the following pairs of points:

10. $(1, 1)$, $(1.1, 1.21)$
11. $(1, 1)$, $(1.01, 1.0201)$
12. $(1, 1)$, $(1+h, 1+k)$
13. $(1, 1)$, $(1+h, 1+2h+h^2)$
14. $(2, 4)$, $(2.1, 4.41)$
15. $(2, 4)$, $(2.01, 4.0401)$

8.2 The meanings of the gradients of lines

In the previous section we saw that the gradient of a line is represented also by the tangent of the angle which the line makes with Ox. There is another meaning which helps to shed more light on the meanings of gradients and which will help us to understand later work better.

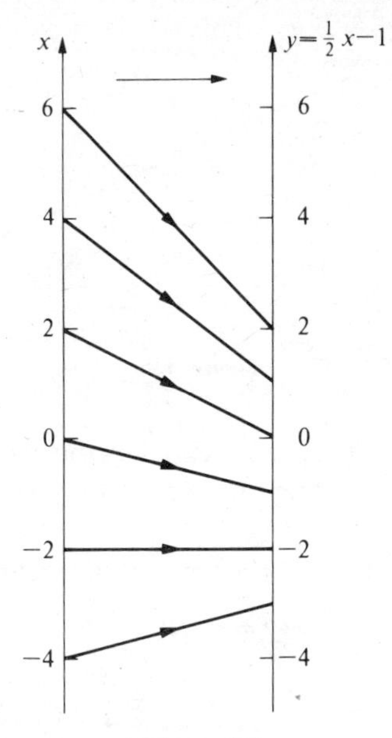

Figure 8.3

Fig. 8.3 represents the details of Fig. 8.1 when the two axes are drawn parallel to one another. It is a mapping diagram. It represents the way in which points on Ox are mapped by the function $y = \frac{1}{2}x - 1$ onto Oy. Point P $(0, -1)$ in Fig. 8.2 relates 0 on Ox to -1 on Oy. Point P_1 $(2, 0)$ in Fig. 8.2 relates 2 on Ox to 0 on Oy. Point P_2 $(4, 1)$ in Fig. 8.2 relates 4 on Ox to 1 on Oy. Point P_3 $(6, 2)$ in Fig. 8.2 relates 6 on Ox to 2 on Oy.

The relationship represented by Fig. 8.3 is called a mapping because one way of looking at the lines in Figs 8.1 and 8.2 is that they map the points of Ox onto Oy. From Fig. 8.3 it can be seen that all intervals marked on Ox are of equal length, 2. The intervals which they map on Oy are also of equal length, but the length is 1. Each interval on the y axis is half the length of the corresponding interval on the x axis. The mapping is said to have a magnification of $\frac{1}{2}$.

Consider another example. The line $y = 3x + 2$ must have a gradient 3 because $m = 3$.

Second, it passes through the point $(0, 2)$: when we put $x = 0$ in the equation, $y = 3 \times 0 + 2 = 2$. It also passes through the point $(1, 5)$: when we put $x = 1$ in the equation, $y = 3 \times 1 + 2 = 5$. The gradient can also be calculated as:

$$\frac{5-2}{1-0} = \frac{3}{1} = 3$$

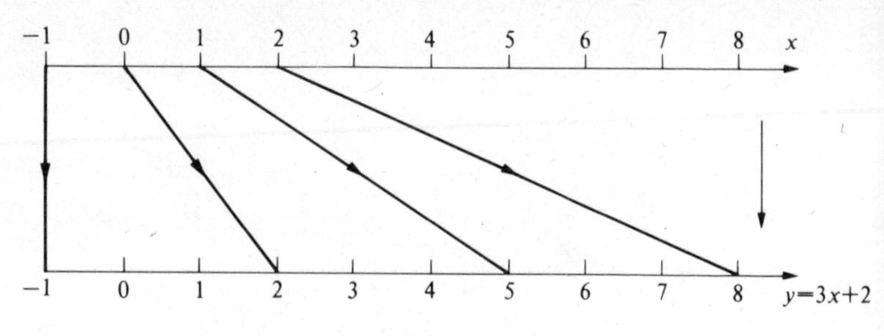

Figure 8.4

Third, the relationship may be represented by the mapping in Fig. 8.4. Each interval of 1 on the axis of x maps onto an interval of 3 on the axis of y. The magnification is 3.

Consider the equation $4y + 3x - 5 = 0$. It represents a line. Rearrange it:

$$4y = -3x + 5$$

$$y = -\frac{3}{4}x + \frac{5}{4}$$

Interpretation 1. $m = -3/4$, so the gradient is $-\frac{3}{4}$.

Interpretation 2. Two points on the line are $(0, \frac{5}{4})$, by substituting $x = 0$, and $(1, \frac{1}{2})$, by substituting $x = 1$. Therefore the gradient $= (\frac{1}{2} - \frac{5}{4})/(1-0)$ $= -\frac{3/4}{1} = -3/4$.

Interpretation 3. Fig. 8.5 gives the mapping relationship. An interval of 1 in the positive sense on Ox maps onto an interval of 3/4 in the negative sense on Oy. The magnification is $-3/4$.

Exercise 8.2

Determine the gradients of the following lines and the magnification of the mappings:

1. $y = 2x - 5$
2. $y = -\frac{1}{4}x + 2$
3. $y + x - 4 = 0$
4. $y - 3x + 6 = 0$
5. $3y - 2x = 7$
6. $2x - y = 4$
7. $5y + 4x - 2 = 0$

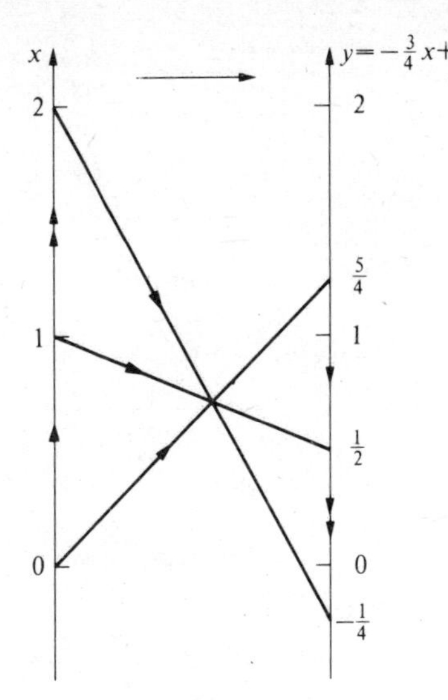

Figure 8.5

8. $3-4x = 2y$
9. $y+ax = c$
10. $ay+bx+c = 0$
11. $ax+by+c = 0$
12. $\frac{x}{2}+\frac{y}{3} = 1$
13. $\frac{x}{a}+\frac{y}{b} = 1$
14. $\frac{ax}{b}+\frac{cy}{d} = 1$

8.3 Gradients of curves

For the time being the second of the above three methods of determining the gradient of a line is going to be the most useful. It will be used to determine the gradients of curves. Fig. 8.6 represents a sketch of the curve

$$y = x^2 \tag{8.2}$$

If we imagined ourselves travelling along the curve from O to K we would appreciate that the further we travelled along the curve the greater the

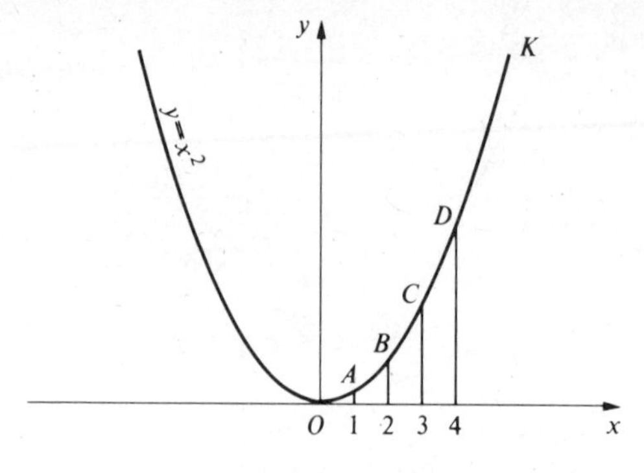

Figure 8.6

gradient would become. In other words, the gradient at $B >$ the gradient at A, the gradient at $C >$ the gradient at B, the gradient at $D >$ the gradient at C, and so on.

How can we measure the gradient of the curve at A, B, C, D, etc.? We turn this problem, which at the moment we cannot answer, into one which we can, namely the gradient of a line. As a preliminary to calculating the gradient of the curve in Fig. 8.6 we determine the average gradients along the sections of the curve.

Average gradients

In Fig. 8.6 the points A, B, C and D divide part of the curve into the sections AB, BC and CD. Travelling along the curve from A to B has exactly the same outcome as travelling from A to B along the straight line AB. We start at A and finish at B.

The co-ordinates of A are $(1, 1)$ and the co-ordinates of B are $(2, 4)$, using equation *8.2*. The gradient of the line AB is, by *8.1*, $\dfrac{4-1}{2-1} = 3$. The average gradient along the section of the curve AB is 3.

The co-ordinates of C are $(3, 9)$ and the co-ordinates of D are $(4, 16)$. The gradient of the line BC is $\dfrac{9-4}{3-2} = 5$. The average gradient along the section of the curve BC is 5.

The average gradient along the section of the curve $CD = \dfrac{16-9}{4-3} = 7$, which equals the gradient of the line CD.

The average gradient of the curve along the section $AC = \dfrac{9-1}{3-1} = \dfrac{8}{2} = 4,$
which is equal to the gradient of the line AC.

Examples

Determine the average gradients of the following curves between the points on the curves with the given values of x.

1. $y = x^2$: between $x = 1$ and $x = 1 + h$.
   When $x = 1$, $y = 1^2 = 1$. Take P_1 to be $(1, 1)$.
   When $x = 1 + h$, $y = (1+h)^2 = 1 + 2h + h^2$. Take P_2 to be $[(1+h), (1+2h+h^2)]$.
   The average gradient of the curve between P_1 and P_2 is the gradient of the line $P_1P_2 = \dfrac{1+2h+h^2-1}{1+h-1} = \dfrac{2h+h^2}{h} = 2 + h.$
2. $y = x^2 + 1$: between $x = 1$ and $x = 1 + h$.
   When $x = 1$, $y = 1^2 + 1 = 2$. Take P_1 to be $(1, 2)$.
   When $x = 1 + h$, $y = (1+h)^2 + 1 = 1 + 2h + h^2 + 1 = 2 + 2h + h^2$.
   Take P_2 to be $[(1+h), (2+2h+h^2)]$.
   The average gradient $= \dfrac{2+2h+h^2-2}{1+h-1} = \dfrac{2h+h^2}{h} = 2 + h.$
3. $y = x^2 + x$: between $x = 1$ and $x = 1 + h$.
   When $x = 1$, $y = 1^2 + 1 = 2$. Take P_1 to be $(1, 2)$.
   When $x = 1 + h$, $y = (1+h)^2 + (1+h) = 1 + 2h + h^2 + 1 + h$
   $= 2 + 3h + h^2$. Take P_2 to be $[(1+h), (2+3h+h^2)]$.
   The average gradient of the curve is the gradient of the line P_1P_2
   $= \dfrac{2+3h+h^2-2}{1+h-1} = \dfrac{3h+h^2}{h} = 3 + h.$

Exercise 8.3

Determine the average gradients of the following curves between the values of x given:

1. $y = x^2$: $x = 1$ and $x = 1.5$
2. $y = x^2$: $x = 1$ and $x = 1.1$
3. $y = x^2$: $x = 1$ and $x = 1.01$
4. $y = x^2$: $x = 2$ and $x = 2.5$
5. $y = x^2$: $x = 2$ and $x = 2.1$

6. $y = x^2$: $x = 2$ and $x = 2.01$
7. $y = x^2 + 1$: $x = 1$ and $x = 2$
8. $y = x^2 + 1$: $x = 1$ and $x = 1.1$
9. $y = x^2 + x$: $x = 1$ and $x = 1.1$
10. $y = x^2 + x$: $x = 1$ and $x = 1.01$
11. $y = x^2$: $x = 2$ and $x = 2 + h$
12. $y = x^2$: $x = 3$ and $x = 3 + h$
13. $y = x^2 + 1$: $x = 2$ and $x = 2 + h$
14. $y = x^2 + 1$: $x = 3$ and $x = 3 + h$
15. $y = x^2 + x$: $x = 2$ and $x = 2 + h$
16. $y = x^2 + x$: $x = 3$ and $x = 3 + h$

Instantaneous gradients

In Fig. 8.6 the gradient of the curve at A is the same as the gradient of the tangent to the curve at A. One method of calculating the gradient of a curve at a point would be to construct an accurate graph, draw an accurate tangent to the curve at the chosen point and then measure the gradient of the tangent. The method is laborious; it is also difficult to achieve accurate results. Instead the method of average gradients is adopted, and this does give accurate results. Fig. 8.7 illustrates the principle adopted. To determine the gradient of the curve at point P, as a first approximation, calculate the average gradient of the section PP_1, where P_1 is a point on the curve close to P. By taking a succession of points closer and closer to P, e.g. P_1, P_2, P_3 etc., and calculating the average gradients of the sections PP_1, PP_2, PP_3 etc., results are obtained which approach steadily closer to the gradient of the tangent PT, at P, to the curve.

In Exercise 8.3 the answers to questions (1), (2) and (3) are 2.5, 2.1 and 2.01. In Example 1 in the preceding examples the gradient is $2 + h$. When P moves closer to P_1, $h \to 0$, and the gradient $2 + h \to 2$. The values 2.5, 2.1 and 2.01 support that conclusion.

We conclude that the instantaneous gradient of $y = x^2$ at $x = 1$ is 2.

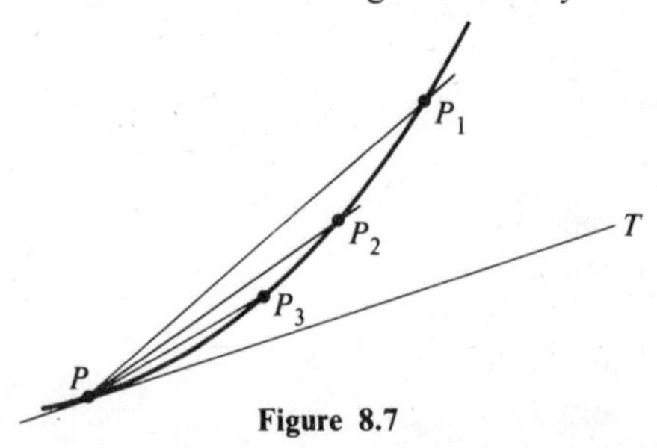

Figure 8.7

Examples

Determine the instantaneous gradients of the following curves at the points indicated.

1. $y = 3x^2$ at the point (1, 3).
   Take the following pairs of points of the curve: (1, 3) and $[(1+h), 3(1+h)^2]$. Then the average gradient is:

$$\frac{3(1+h)^2 - 3}{1 + h - 1}$$

$$= \frac{3 + 6h + 3h^2 - 3}{h} = \frac{6h + 3h^2}{h} = 6 + 3h$$

   As the second point moves closer to the first $h \to 0$. The gradient $\to 6$. This is the instantaneous gradient at (1, 3).
2. Determine the instantaneous rate of change for the curve $y = -4x^2$ at the point where $x = 3$.
   Step 1. On the curve when $x = 3$, $y = -4x^2 = -4 \times 9 = -36$.
   Step 2. Take P_1 to be (3, −36) and P_2 to be where $x = 3 + h$.
   Step 3. On the curve when $x = 3 + h$, $y = -4(3+h)^2$
   $= -4(9 + 6h + h^2) = -36 - 24h - 4h^2$
   Step 4. P_2 is $[(3+h), (-36 - 24h - 4h^2)]$.
   Step 5. The average gradient of the curve between P_1 and P_2 is:

$$\frac{-36 - 24h - 4h^2 - (-36)}{3 + h - 3} = \frac{-24h - 4h^2}{h} = -24 - 4h$$

   Step 6. Assume $P_2 \to P_1$, i.e. $h \to 0$.
   Step 7. Then the average gradient $\to -24$.
   Step 8. The instantaneous gradient of the curve (rate of change) at $P_1 = -24$.

Exercise 8.4

Determine the instantaneous rates of change of the following functions for the given values of x, i.e. find the gradients of the associated curves at the points whose co-ordinates are given.

1. $2x^2$: where $x = 2$
2. $2x^2$: where $x = 3$
3. $2x^2$: where $x = -1$
4. $5x^2$: where $x = 1$

5. $5x^2$: where $x = 2$
6. $-3x^2$: where $x = 1$
7. $-4x^2$: where $x = 2$
8. $-4x^2$: where $x = -1$
9. $x^2 + 3$: where $x = 1$
10. $x^2 + x$: where $x = 1$
11. $x^2 + 2x$: where $x = 1$
12. $x^2 - 2x$: where $x = 1$
13. $x^2 + 2x$: where $x = -1$
14. $x^2 + x + 2$: where $x = 1$

8.4 Rate of change function

The methods adopted so far enable us to determine the gradients a specified points of a curve, i.e. to calculate the values of the rates of chang of functions for particular values of x.

For each point on a given curve a separate calculation has to be made. I would be an advantage if a rule could be discovered for each curve whicl enabled the gradient at any point of the curve to be written down. Fig. 8. and the following explanation represent the principles already established These we shall adapt to determine rules.

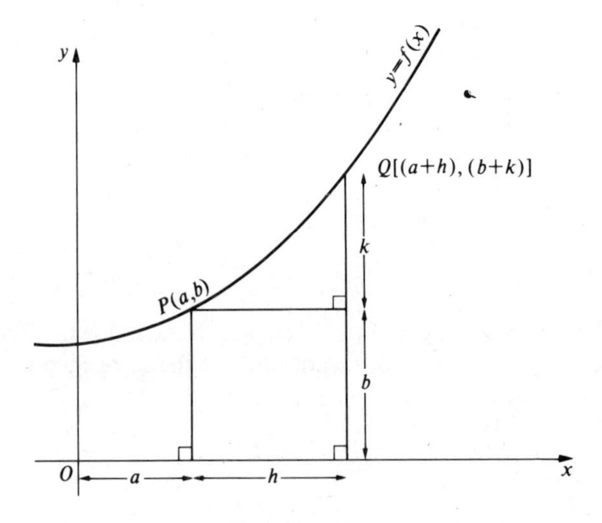

Figure 8.8

1. On the curve $y = f(x)$, P is a point with co-ordinates (a, b).
2. Q is a point on the same curve close to P. Its co-ordinates are $[(a+h), (b+k)]$.
3. Movement from P to Q along the curve produces a change in x of h (called an increment).
4. The same movement produces an increment in y of k.
5. The average gradient of the curve between P and Q is $\dfrac{(\not{b}+k)-\not{b}}{(\not{a}+h)-\not{a}} = \dfrac{k}{h}$
6. The instantaneous gradient of the curve at P is obtained by allowing h to tend to 0.
7. The a and b have always been particular numerical values.

In order that general rules may be determined a and b must have general values for a curve.

The following examples illustrate the modification to the above method.

Examples

1. Determine a rule for obtaining the gradient at any point of the curve $y = x^2$.
   Fig. 8.9 is a sketch of the curve. The x co-ordinate of P is x. So the y co-ordinate of P is found by substituting that value of x in the equation $y = x^2$. The co-ordinates of P are (x, x^2).

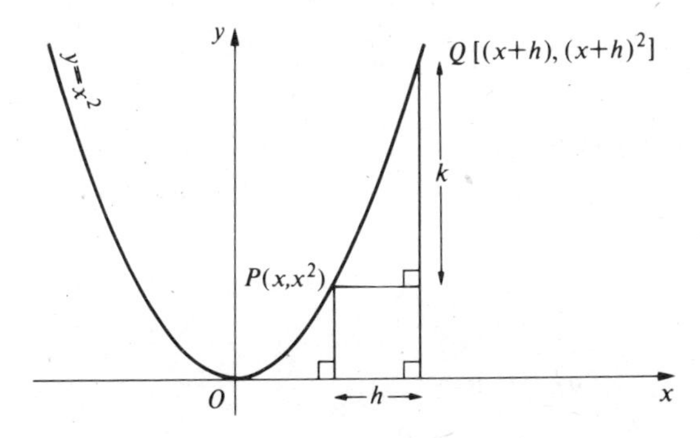

Figure 8.9

The x co-ordinate of Q is $(x+h)$. So the y co-ordinate of Q is found by substituting that x value in $y = x^2$. The y co-ordinate of Q is $(x+h)^2$. The co-ordinates of Q are $[(x+h), (x+h)^2]$.
The increment in x from P to Q is h.
The increment in y from P to Q is k.

The average gradient of the curve between P and Q is:

$$\frac{(x+h)^2-x^2}{x+h-x}=\frac{x^2+2xh+h^2-x^2}{h}=2x+h\text{, i.e.}$$

$$\frac{k}{h}=2x+h$$

As $Q \to P$, $h \to 0$, the average gradient $\to 2x$. The instantaneous gradient at P is $2x$. This instantaneous gradient is itself a function of x. It is called the derived function of x^2 (the original function for the curve). By substituting particular values of x in $2x$ the gradients at particular points may be obtained.

2. Determine a rule for obtaining the derived function of $y = ax^2$.

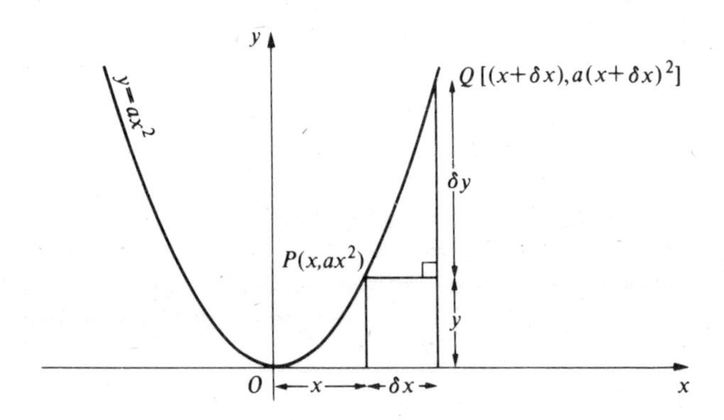

Figure 8.10

In Fig. 8.10 a new notation is introduced. The increment h is written δx (meaning a little bit of x, not $\delta \times x$). The increment k is written δy. The average gradient is:

$$\frac{\delta y}{\delta x}=\frac{a(x+\delta x)^2-ax^2}{x+\delta x-x}=\frac{ax^2+2ax\,.\,\delta x+a\,.\,\delta x^2-ax^2}{x}$$

$$=2ax+a\,.\,\delta x$$

$$\text{i.e.}\quad \frac{\delta y}{\delta x}=2ax+a\,.\,\delta x$$

As $\delta x \to 0$, $\dfrac{\delta y}{\delta x} \to 2ax$

The derived function of ax^2 is $2ax$. Where $\dfrac{\delta y}{\delta x}$ tends to a limit, as above, that limit is said to be $\frac{dy}{dx}$. In other words:

$$\frac{dy}{dx} = \underset{\delta x \to 0}{\text{Lt}} \left(\frac{\delta y}{\delta x}\right)$$

3. Obtain the derived function of $y = ax^2 + bx + c$.

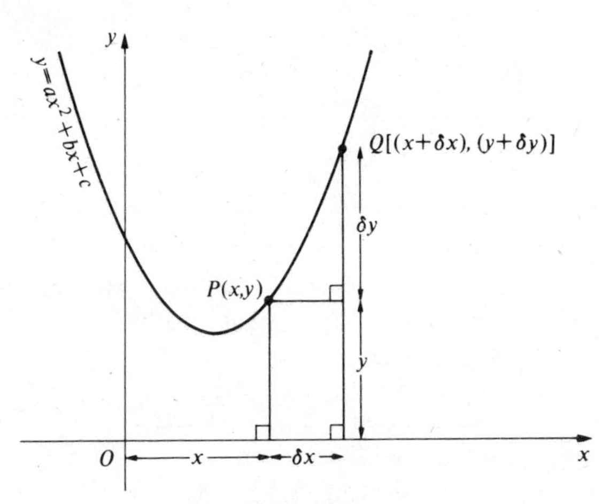

Figure 8.11

In Fig. 8.11 the y co-ordinate of Q is $a(x+\delta x)^2 + b(x+\delta x) + c$. The average gradient of PQ is:

$$\frac{\delta y}{\delta x} = \frac{a(x+\delta x)^2 + b(x+\delta x) + c - (ax^2 + bx + c)}{\not{x} + \delta x - \not{x}}$$

$$= \frac{1}{\delta x}\left[ax^2 + 2ax\,.\,\delta x + a\,.\,\delta x^2 + bx + b\,.\,\delta x + c - ax^2 - bx - c\right]$$

$$= 2ax + b + a\,.\,\delta x$$

As $\delta x \to 0$, $\frac{\delta y}{\delta x} \to 2ax + b$

Then $\frac{dy}{dx} = 2ax + b$.

Exercise 8.5

Determine the derived functions of the following functions:

1. $6x^2$
2. $-4x^2$
3. $\frac{2}{3}x^2$
4. x^2+x
5. x^2+3x
6. x^2-x
7. ax^2-bx
8. $\frac{a}{b}x^2$
9. px^2-qx-r

Examples

1. Use the derived function in Example 1 above to calculate the gradients of the curve $y = x^2$ for the following values of x: $x = 0, -2, 3, 1.5, -\frac{1}{3}$.
   The derived function is $2x$.
   When $x = 0$, the gradient $= 2 \times 0 = 0$.
   When $x = -2$, the gradient $= 2 \times -2 = -4$.
   When $x = 3$, the gradient $= 2 \times 3 = 6$.
   When $x = 1.5$, the gradient $= 2 \times 1.5 = 3$.
   When $x = -\frac{1}{3}$, the gradient $= 2 \times -\frac{1}{3} = -2/3$.
2. Use the derived function in Example 2 above to calculate the gradients of the curve $y = ax^2$ for the following values of x:$x = 0$, $2\frac{1}{4}$, $-3\frac{1}{2}$, $5\frac{1}{3}$, b, $-c$, p/q.
   The derived function is $2ax$.
   When $x = 0$, the gradient $= 2a \times 0 = 0$.
   When $x = 2\frac{1}{4}$, the gradient $= 2a \times \frac{9}{4} = 9a/2$.
   When $x = -3\frac{1}{2}$, the gradient $= 2a \times -\frac{7}{2} = -7a$.
   When $x = 5\frac{1}{3}$, the gradient $= 2a \times \frac{16}{3} = \dfrac{32a}{3}$
   When $x = b$, the gradient $= 2a \times b = 2ab$.
   When $x = -c$, the gradient $= 2a \times -c = -2ac$.
   When $x = p/q$, the gradient $= 2a \times \dfrac{p}{q} = 2ap/q$.
3. Use the derived function of Example 3 above to calculate the gradients of the curve $y = ax^2 + bx + c$ for the following values of x: $x = 0, 4, -2, 1\frac{1}{2}, p$.
   The derived function is $2ax + b$.
   When $x = 0$, the gradient $= 2a \times 0 + b = b$.
   When $x = 4$, the gradient $= 2a \times 4 + b = 8a + b$.

When $x = -2$, the gradient $= 2a \times -2 + b = -4a + b$.
When $x = 1\frac{1}{2}$, the gradient $= 2a \times 1\frac{1}{2} + b = 3a + b$.
When $x = p$, the gradient $= 2a \times p + b = 2ap + b$.

Exercise 8.6

Use the appropriate derived functions from Exercise 8.5 to calculate the gradients of the following curves at the values of x indicated.

1. $y = 6x^2$: $x = 0, 1, -2, 3, -4, a$
2. $y = -4x^2$: $x = 0, \frac{1}{2}, -\frac{1}{4}, \frac{1}{3}, -2, b$
3. $y = \frac{2}{3}x^2$: $x = -2, -3, 4, 5, -\frac{1}{4}, a$
4. $y = x^2 + x$: $x = 0, -2, 4, -6, 8, a$
5. $y = x^2 + 3x$: $x = 0, 1, -2, 2/3, -5/4$

8.5 Gradients at maximum and minimum points

Fig. 8.12 is a sketch of a curve. The point A is a minimum point. The rate of change of the function at A = the gradient of the curve at $A = 0$.

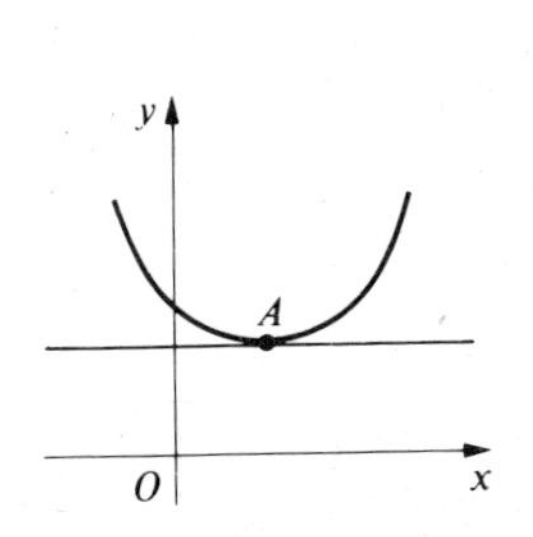

Figure 8.12

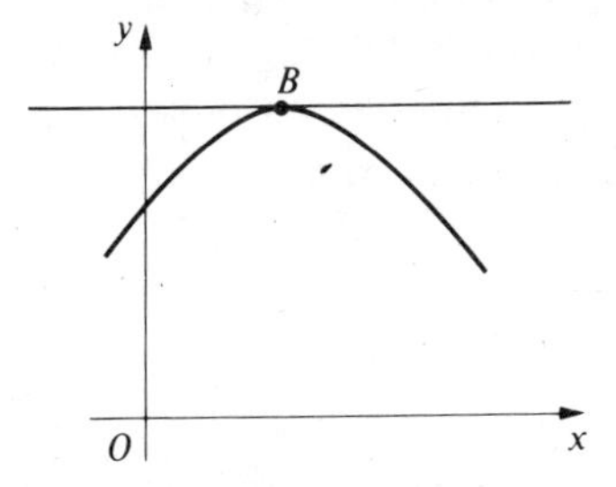

Figure 8.13

Fig. 8.13 is a sketch of a curve. The point B is a maximum point. The rate of change of the function at B = the gradient of the curve at $B = 0$.

9

Differentiation of Simple Algebraic and Trigonometric Functions

Differentiation is an alternative term for the operation of determining the derived function of a given function. Various alternative standard symbols are used to denote the derived function. When $y = \mathrm{f}(x)$ the derived function may be written:

$$\frac{\mathrm{d}y}{\mathrm{d}x} \quad \text{or} \quad \mathrm{d}y/\mathrm{d}x \quad \text{or} \quad \frac{\mathrm{d}}{\mathrm{d}x}[\mathrm{f}(x)] \quad \text{or} \quad \mathrm{f}'(x)$$

We say that we have differentiated $\mathrm{f}(x)$ with respect to x, i.e. w.r.t. x.

To determine the derivatives of a wide variety of functions a set of rules is required. The following examples illustrate the method of determining the rules for a number of simple algebraic functions. The methods are those developed in Chapter 8.

Examples

Determine the derived functions of ax^n in the following cases:

1. When $n = 0$.

$$y = a.x^0 = a \text{ (constant)} \qquad \text{(i)}$$

Take P_1 to be (x, y). From (i), $y = a$, for all values of x.
Then $P_1 \equiv (x, a)$.
Take P_2 to be $[(x+\delta x), (y+\delta y)]$. From (i), $y+\delta y = a$.
Then $P_2 \equiv [(x+\delta x), a]$.

The average gradient between P_1 and P_2 on the curve (i) is:

$$\frac{a-a}{x+\delta x-x}=\frac{0}{\delta x}=0$$

The derived function of a is 0.
Note that $y = a$ is the equation of a line parallel to Ox, i.e. of gradient 0.

2. When $n = 1$.

$$y = ax \qquad \text{(ii)}$$

Take P_1 to be (x, y). From (ii), $y = ax$.
Then $P_1 \equiv (x, ax)$.
Take P_2 to be $[(x+\delta x), (y+\delta y)]$. From (ii), $y+\delta y = a(x+\delta x)$.
Then $P_2 \equiv [(x+\delta x), a(x+\delta x)]$.
The average gradient between P_1 and P_2 on the curve (ii) is:

$$\frac{a(x+\delta x)-ax}{x+\delta x-x}=\frac{ax+a.\delta x-ax}{\delta x}=\frac{a.\delta x}{\delta x}$$

$$= a \text{ (constant)}$$

As $\delta x \to 0$ the average gradient stays at a. The derived function of ax is a, i.e. $\frac{\mathrm{d}}{\mathrm{d}x}(ax) = a$.

Note that $y = ax$ is the equation of a line whose gradient is a.

3. When $n = 2$.

$$y = ax^2 \qquad \text{(iii)}$$

Take P_1 to be (x, y). From (iii), $y = ax^2$.
Then $P_1 \equiv (x, ax^2)$.
Take P_2 to be $[(x+\delta x), (y+\delta y)]$. From (iii), $y+\delta y = a(x+\delta x)^2$.
Then $P_2 \equiv [(x+\delta x), a(x+\delta x)^2]$.
The average gradient between P_1 and P_2 on the curve (iii) is:

$$\frac{a(x+\delta x)^2-ax^2}{x+\delta x-x}=\frac{ax^2+2ax.\delta x+a.\delta x^2-ax^2}{\delta x}$$

$$= 2ax + a.\delta x$$

As $\delta x \to 0$ the average gradient $\to 2ax$. The derived function of ax^2 is $2ax$, i.e. $\frac{\mathrm{d}}{\mathrm{d}x}(ax^2) = 2ax$.

4. When $n = 3$.

$$y = ax^3 \qquad \text{(iv)}$$

Take P_1 to be (x, y). From (iv), $y = ax^3$.
Then $P_1 \equiv (x, ax^3)$.
Take P_2 to be $[(x+\delta x), (y+\delta y)]$. From (iv), $y+\delta y = a(x+\delta x)^3$.
Then $P_2 \equiv [(x+\delta x), a(x+\delta x)^3]$.
The average gradient between P_1 and P_2 is:

$$\frac{a(x+\delta x)^3 - ax^3}{\cancel{x}+\delta x-\cancel{x}} = \frac{\cancel{ax^3}+3ax^2.\delta x+3ax.\delta x^2+\delta x^3-\cancel{ax^3}}{\delta x}$$

$$= 3ax^2 + 3ax.\delta x + \delta x^2$$

As $\delta x \to 0$ the average gradient $\to 3ax^2$. The derived function of ax^3 is $3ax^2$, i.e. $\frac{d}{dx}(ax^3) = 3ax^2$.

5. Determine the derived function of:

$$y = ax^3 + bx^2 + cx + d \qquad \text{(v)}$$

Note: the derived functions of the separate terms of (v) are $3ax^2$, $2bx$, c and 0 (using (iv), (iii), (ii) and (i) respectively).
Take P_1 to be (x, y). From (v), $y = ax^3 + bx^2 + cx + d$.
Then $P_1 \equiv (x, ax^3 + bx^2 + cx + d)$.
Take P_2 to be $[(x+\delta x), (y+\delta y)]$. From (v), $y = a(x+\delta x)^3 + b(x+\delta x)^2 + c(x+\delta x)+d$.
Then $P_2 \equiv [(x+\delta x), a(x+\delta x)^3 + b(x+\delta x)^2 + c(x+\delta x)+d]$.
The average gradient between P_1 and P_2 is:

$$\frac{a(x+\delta x)^3 + b(x+\delta x)^2 + c(x+\delta x)+d-(ax^3+bx^2+cx+d)}{\cancel{x}+\delta x-\cancel{x}}$$

$$= \frac{\cancel{ax^3}+3ax^2.\delta x+3ax.\delta x^2+\delta x^3+\cancel{bx^2}+2bx.\delta x+b.\delta x^2+\cancel{cx} +c\,\delta x+\cancel{d}-\cancel{ax^3}-\cancel{bx^2}-\cancel{cx}-\cancel{d}}{\delta x}$$

$$= 3ax^2 + 3ax.\delta x + \delta x^2 + 2bx + b.\delta x + c$$

$$= (3ax^2 + 2bx + c) + (3ax + b).\delta x + \delta x^2$$

As $\delta x \to 0$ the average gradient $\to 3ax^2 + 2bx + c$

The derived function of $ax^3 + bx^2 + cx + d$ is $3ax^2 + 2bx + c$, i.e.

$$\frac{d}{dx}(ax^3 + bx^2 + cx + d) = 3ax^2 + 2bx + c$$

Note: this is the sum of the derived functions of the separate terms, i.e.

$$\frac{d}{dx}(ax^3 + bx^2 + cx + d) = \frac{d}{dx}(ax^3) + \frac{d}{dx}(bx^2) + \frac{d}{dx}(cx) + \frac{d}{dx}(d) \qquad 9.1$$

9.1 is a special case of a much more general principle: when f(x), g(x), h(x), etc. are functions of x which are differentiable (i.e. it is possible to differentiate them), then:

$$\frac{d}{dx}[f(x)+g(x)+h(x)+\ldots] = \frac{d}{dx}\{f(x)\} + \frac{d}{dx}\{g(x)\} + \frac{d}{dx}\{h(x)\} + \ldots] \qquad 9.2$$

Examples

The following are special cases of the examples above.

1. In Example 1 above put $a = 1$.

   Then $y = 1$; $dy/dx = 0$, or $\frac{d}{dx}(1) = 0$.

2. In Example 2 above put $a = 1$.

   Then $y = x$; $dy/dx = 1$, or $\frac{d}{dx}(x) = 1$.

   This means that $\frac{d}{dx}(ax) = a = a \times 1 = a \times \frac{d}{dx}(x)$.

   That is $\frac{d}{dx}(ax) = a.\frac{d}{dx}(x)$.

3. In Example 3 above put $a = 1$.

   Then $y = x^2$; $dy/dx = 2x$, i.e. $\frac{d}{dx}(x^2) = 2x$.

   Therefore $\frac{d}{dx}(ax^2) = a \times 2x = a \cdot \frac{d}{dx}(x^2)$.

4. In Example 4 above put $a = 1$.

   Then $y = x^3$; $dy/dx = 3x^2$, i.e. $\frac{d}{dx}(x^3) = 3x^2$.

   Therefore $\frac{d}{dx}(ax^3) = 3ax^2 = a \times 3x^2 = a \cdot \frac{d}{dx}(x^3)$.

All four of the examples above bear out a relationship of the form:

$$\frac{d}{dx}(ax^n) = a.\frac{d}{dx}(x^n) \qquad 9.3$$

In fact, more generally:

$$\frac{d}{dx}[a.f(x)] = a.\frac{d}{dx}[f(x)], \text{ where } a \text{ is constant} \qquad 9.4$$

The two laws, *9.2* and *9.4*, are two of the most important laws for derivatives which we shall encounter. They indicate that the operation $\frac{d}{dx}$ is a linear operator.

9.1 The derivative of x^n w.r.t. x

Table 9.1

$f(x)$	$f'(x) = \frac{d}{dx} f(x)$
$1 = x^0$	0
$x = x^1$	$1 = 1 \cdot x^0$
x^2	$2x = 2 \cdot x^1$
x^3	$3x^2$
x^4	$4x^3$
x^5	$5x^4$
x^6	$6x^5$

Table 9.1 summarizes the main details of the examples above. The first four lines of Table 9.1 come from the information obtained above. A study of these results suggests a pattern. When that pattern is put into operation a wider range of $f(x)$ can be included in the lines following. Put x^4 under $f(x)$ in line 5. The pattern suggests that the entry under $f'(x)$ on that line is $4x^3$. Put x^5 under $f(x)$ in line 6. The entry under $f'(x)$ is $5x^4$. In fact those results are correct. No attempt will be made at this stage to prove them. The law which covers all such cases is:

$$\frac{d}{dx}(x^n) = n.x^{n-1} \qquad 9.5$$

Law *9.5* is true not only when n is a positive integer. It is true for all real values of n.

Consequently we have at our disposal three laws for determining derived functions of simple algebraic expressions. They are *9.2*, *9.4* and *9.5*. Of these, *9.2* and *9.4* are general principles which can be applied to any functions, not only to algebraic expressions. Law *9.5* is a special law often called a standard form which is applicable only to certain algebraic functions.

Examples

1. Differentiate $15x^7$ w.r.t. x and calculate its value when $x = 2$.

$$\frac{d}{dx}(15x^7) = 15.\frac{d}{dx}(x^7) \quad \text{(by } \textit{9.4}\text{)}$$
$$= 15 \times 7x^6 \quad \text{(by } \textit{9.5}\text{)}$$
$$= 105x^6$$

When $x = 2$, the derivative $= 105 \times 2^6 = 105 \times 64 = 6720$.

2. Differentiate $\frac{13}{5x^3}$ w.r.t. x and calculate its value when $x = -\frac{1}{3}$.

Step 1. Write $\frac{13}{5} \times \frac{1}{x^3} = \frac{13}{5}x^{-3}$

Step 2.

$$\frac{d}{dx}\left[\frac{13}{5}x^{-3}\right] = \frac{13}{5}.\frac{d}{dx}(x^{-3}) \quad \text{(by } \textit{9.4}\text{)}$$
$$= \frac{13}{5} \times -3.x^{-4} \quad \text{(by } \textit{9.5}\text{)}$$
$$= -\frac{39}{5}.x^{-4} = -\frac{39}{5x^4}$$

Step 3. When $x = -\frac{1}{3}$, the derivative $= -\frac{39}{5(-\frac{1}{3})^4} = -\frac{39}{5}.3^4$

$= -\frac{39}{5} \times 81 = -\frac{3159}{5} = -631.8$.

3. Differentiate $\frac{3}{5}\sqrt{x}$ w.r.t. x and calculate the value of the derivative when $x = 12\frac{1}{4}$.

Step 1. Write $\frac{3}{5}\sqrt{x} = \frac{3}{5}.x^{\frac{1}{2}}$

Step 2.

$$\frac{d}{dx}\left[\frac{3}{5}.x^{\frac{1}{2}}\right] = \frac{3}{5}.\frac{d}{dx}(x^{\frac{1}{2}}) \quad \text{(by } \textit{9.4}\text{)}$$
$$= \frac{3}{5} \times \tfrac{1}{2}.x^{-\frac{1}{2}} \quad \text{(by } \textit{9.5}\text{)}$$
$$= \frac{3}{10}.\frac{1}{x^{\frac{1}{2}}} = \frac{3}{10\sqrt{x}}.$$

Step 3. When $x = 12\frac{1}{4}$, i.e. 49/4, the derivative $= \dfrac{3}{10\sqrt{\dfrac{49}{4}}} = \dfrac{3}{10(7/2)}$

$= \dfrac{3}{35}$

4. Differentiate $y = 11x^5 + 3x^2 + 6/x - 2/x^4$ w.r.t. x and calculate the value of dy/dx when $x = -1$.
Step 1. Write $y = 11x^5 + 3x^2 + 6x^{-1} + (-2).x^{-4}$.
Step 2.

$$\frac{dy}{dx} = \frac{d}{dx}(11x^5) + \frac{d}{dx}(3x^2) + \frac{d}{dx}(6x^{-1}) + \frac{d}{dx}(-2x^{-4}) \quad \text{(by 9.2)}$$

$$= 11.\frac{d}{dx}(x^5) + 3.\frac{d}{dx}(x^2) + 6.\frac{d}{dx}(x^{-1}) + (-2).\frac{d}{dx}(x^{-4}) \quad \text{(by 9.4)}$$

$$= 11 \times 5x^4 + 3 \times 2x^1 + 6(-1)x^{-2} + (-2)(-4)x^{-5} \quad \text{(by 9.5)}$$

$$= 55x^4 + 6x - \frac{6}{x^2} + \frac{8}{x^5}$$

Step 3. When $x = -1$, $dy/dx = 55(-1)^4 + 6(-1) - \dfrac{6}{(-1)^2} + \dfrac{8}{(-1)^5}$

$= 55 - 6 - 6 - 8 = 55 - 20 = 35.$

5. Differentiate $y = 5x\sqrt{x} - \dfrac{3}{\sqrt{x}} + \dfrac{2}{x\sqrt{x}}$ w.r.t. x and calculate the value of dy/dx when $x = 4$.
Step 1. Write $y = 5x.x^{\frac{1}{2}} - 3.x^{-\frac{1}{2}} + 2.x^{-3/2}$
Step 2.

$$dy/dx = \frac{d}{dx}[5x^{3/2} - 3x^{-\frac{1}{2}} + 2x^{-3/2}]$$

$$= \frac{d}{dx}(5x^{3/2}) + \frac{d}{dx}(-3x^{-\frac{1}{2}}) + \frac{d}{dx}(2x^{-3/2}) \quad \text{(by 9.2)}$$

$$= 5\frac{d}{dx}(x^{3/2}) - 3\frac{d}{dx}(x^{-\frac{1}{2}}) + 2\frac{d}{dx}(x^{-3/2}) \quad \text{(by 9.4)}$$

$$= 5.\frac{3}{2}.x^{\frac{1}{2}} - 3(-\frac{1}{2}).x^{-3/2} + 2\left(-\frac{3}{2}\right).x^{-5/2} \quad \text{(by 9.5)}$$

$$= \frac{15}{2}\sqrt{x} + \frac{3}{2}.\frac{1}{x^{3/2}} - 3.\frac{1}{x^{5/2}}$$

$$= \frac{15}{2}\sqrt{x} + \frac{3}{2x\sqrt{x}} - \frac{3}{x^2\sqrt{x}}$$

Step 3. When $x = 4, dy/dx = \frac{15}{2} \times 2 + \frac{3}{2.4.2} - \frac{3}{16.2} = 15 + \frac{3}{16} - \frac{3}{32}$

$= 15 + \frac{3}{32} = 15\frac{3}{32}$.

6. Differentiate $y = (5 + x - 2x^2)(3 + 4x)$ w.r.t. x and calculate the gradient when $x = -3/4$.

Step 1. Write:

$$\begin{aligned} y &= (5 + x - 2x^2).3 + (5 + x - 2x^2).4x \\ &= 15 + 3x - 6x^2 + 20x + 4x^2 - 8x^3 \\ &= 15 + 23x - 2x^2 - 8x^3 \end{aligned}$$

Step 2.

$$\begin{aligned} dy/dx &= \frac{d}{dx}(15 + 23x - 2x^2 - 8x^3) \\ &= \frac{d}{dx}(15) + \frac{d}{dx}(23x) + \frac{d}{dx}(-2x^2) + \frac{d}{dx}(-8x^3) \quad \text{(by 9.2)} \\ &= 15\frac{d}{dx}(1) + 23\frac{d}{dx}(x) - 2\frac{d}{dx}(x^2) - 8\frac{d}{dx}(x^3) \quad \text{(by 9.4)} \\ &= 15 \times 0 + 23 \times 1 - 2 \times 2x - 8 \times 3x^2 \quad \text{(by 9.5)} \\ &= 23 - 4x - 24x^2 \end{aligned}$$

Step 3. When $x = -3/4$, $dy/dx = 23 - 4(-3/4) - 24(-\frac{3}{4})^2$

$= 23 + 3 - 24 \times \frac{9}{16} = 26 - 13\frac{1}{2} = 12\frac{1}{2}$.

7. Differentiate $y = \frac{(3 + 5x - 6x^2)}{6x}$ w.r.t. x and calculate the value of dy/dx when $x = 2\frac{1}{2}$.

Step 1. Write:

$$\begin{aligned} y &= \frac{3}{6x} + \frac{5x}{6x} - \frac{6x^2}{6x} \\ &= \frac{1}{2x} + \frac{5}{6} - x \\ &= \frac{1}{2}.x^{-1} + \frac{5}{6} - x \end{aligned}$$

Step 2.

$$dy/dx = \frac{d}{dx}\left[\tfrac{1}{2}.x^{-1} + \frac{5}{6} - x\right]$$

$$= \frac{d}{dx}(\tfrac{1}{2}.x^{-1}) + \frac{d}{dx}\left(\frac{5}{6}\right) + \frac{d}{dx}(-x) \text{ (by 9.2)}$$

$$= \tfrac{1}{2}.\frac{d}{dx}(x^{-1}) + \frac{5}{6}\frac{d}{dx}(1) - \frac{d}{dx}(x) \quad \text{(by 9.4)}$$

$$= \tfrac{1}{2}(-1).x^{-2} + \frac{5}{6} \times 0 - 1 \quad \text{(by 9.5)}$$

$$= -\frac{1}{2x^2} - 1$$

Step 3. When $x = 2\tfrac{1}{2}$, $dy/dx = -\dfrac{1}{2\left(\dfrac{5}{2}\right)^2} - 1 = -\dfrac{4}{2 \times 25} - 1$

$$= -\frac{2}{25} - 1 = -1\frac{2}{25}$$

8. Given $y = \dfrac{(x+2)^3}{\sqrt{x}}$ determine dy/dx and calculate its value when $x = \tfrac{1}{4}$.

Step 1. Write $y = \dfrac{x^3 + 6x^2 + 12x + 8}{\sqrt{x}} = x^{5/2} + 6x^{3/2} + 12x^{\frac{1}{2}} + 8x^{-\frac{1}{2}}$

Step 2.

$$\frac{dy}{dx} = \frac{d}{dx}(x^{5/2}) + \frac{d}{dx}(6x^{3/2}) + \frac{d}{dx}(12x^{\frac{1}{2}}) + \frac{d}{dx}(8x^{-\frac{1}{2}}) \quad \text{(by 9.2)}$$

$$= \frac{d}{dx}(x^{5/2}) + 6\frac{d}{dx}(x^{3/2}) + 12\frac{d}{dx}(x^{\frac{1}{2}}) + 8\frac{d}{dx}(x^{-\frac{1}{2}}) \quad \text{(by 9.4)}$$

$$= \tfrac{5}{2}x^{3/2} + 6.\tfrac{3}{2}.x^{\frac{1}{2}} + 12.\tfrac{1}{2}.x^{-\frac{1}{2}} + 8.-\tfrac{1}{2}.x^{-3/2} \quad \text{(by 9.5)}$$

$$= \frac{5}{2}x\sqrt{x} + 9\sqrt{x} + \frac{6}{\sqrt{x}} - \frac{4}{x\sqrt{x}}$$

Step 3. When $x = \tfrac{1}{4}$, $dy/dx = \tfrac{5}{2} \times \tfrac{1}{4} \times \tfrac{1}{2} + 9 \times \tfrac{1}{2} + 6 \times 2 - 4 \times 4 \times 2$

$$= \frac{5}{16} + 4\tfrac{1}{2} + 12 - 32 = -16 + \frac{13}{16} = -15\frac{3}{16}$$

Exercise 9.1

Determine the derivatives of the following functions w.r.t. the independent variables:

1. x^8
2. x^{10}
3. x^{12}
4. x^{-3}
5. x^{-8}
6. $1/x^2$
7. $1/x^4$
8. x^{-6}
9. $x^{\frac{1}{2}}$
10. $x^{3/2}$
11. $x^{5/2}$
12. $x^{-\frac{1}{2}}$
13. $x^{-3/2}$
14. $x^{-5/4}$
15. $\sqrt{x}$
16. $x\sqrt{x}$
17. $\dfrac{1}{u^2\sqrt{u}}$
18. $\dfrac{1}{\sqrt[4]{u^3}}$
19. $\sqrt[3]{y^2}$
20. $\sqrt{y^3}$
21. $11x^2$
22. $-\frac{2}{3}x^4$
23. $\dfrac{15}{16x}$
24. $\dfrac{6}{5x^2}$
25. $\frac{3}{4}\sqrt{x}$
26. $\dfrac{7}{16}x^{8/7}$
27. $-\dfrac{7}{3x^3\sqrt{x}}$
28. x^2+5x+2
29. $3x^2-7x-8$

30. $4x+2+\dfrac{3}{x}$

31. $11x-1-\dfrac{2}{x^2}$

32. $3x^2+5x-\dfrac{9}{x^2}$

Determine the derivatives of the following functions w.r.t. the independent variables and calculate the values of those derivatives for the indicated values of the variables.

33. $(x-1)^2$: $x=3$
34. $(2x+5)^2$: $x=-1$
35. $(4x+3)(x-7)$: $x=1$
36. $(x^2+2)(3x-5)$: $x=-2$
37. $(2x-1)^3$: $x=2$
38. $\dfrac{(x+1)}{x}$: $x=1$
39. $\dfrac{4x+5}{3x}$: $x=2$
40. $\dfrac{6x^2-3x+7}{5x}$: $x=-3$
41. $\dfrac{(x+2)^2}{5x}$: $x=\frac{1}{2}$
42. $\dfrac{(3u^2-1)(2u+5)}{4u^2}$: $u=-1$
43. $\dfrac{\sqrt{x}+1}{x}$: $x=4$
44. $\dfrac{(\sqrt{x}+2)(x-3)}{x^2}$: $x=1$
45. $8x^3-5x^2+11x-6$: $x=3$
46. $(2x^3-7x-1)(4x+1)$: $x=2$
47. $\dfrac{3}{x}(7x^4-5x^3-9x^2+12x-13)$: $x=2$
48. $\dfrac{2}{x^2}\left(12x^3-15x+\dfrac{7}{x}-\dfrac{5}{x^3}\right)$: $x=-1$
49. $\left(\dfrac{5}{x}+2x\right)\left(\dfrac{1}{x^2}-3x^2\right)$: $x=-2$
50. $3x(4x+1)(2x-3)(5x+4)$: $x=2$

9.2 The derivatives of sin x and cos x w.r.t. x

Fig. 9.1 is a sketch of a graphical method of representing $\sin x$ where x is an angle measured in radians. The circle has a unit radius. The RHS of Fig. 9.1 is the sketch for the graph of $y = \sin x$. The point L on the circle determines point P on the sine curve. Point M on the circle determines point Q on the sine curve. The co-ordinates of P are (x, y); the co-ordinates of Q are $[(x+\delta x), (y+\delta y)]$.

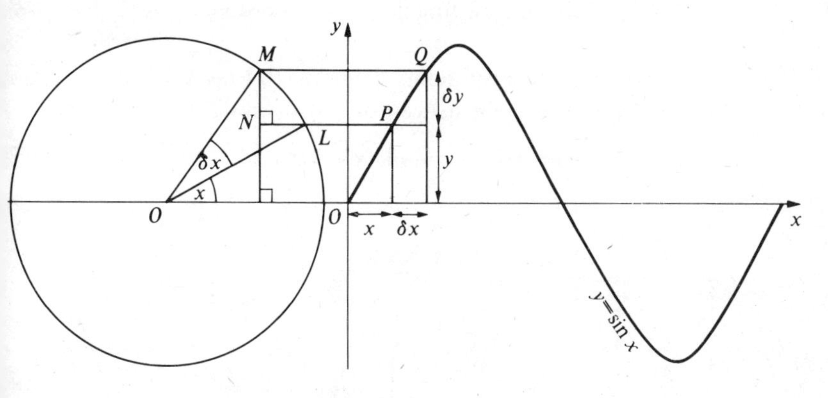

Figure 9.1

The average gradient of the sine curve between P and Q is $\dfrac{\delta y}{\delta x}$.

LN is horizontal and MN vertical. They meet at N. OL makes an angle x with the horizontal. Angle $LOM = \delta x$.

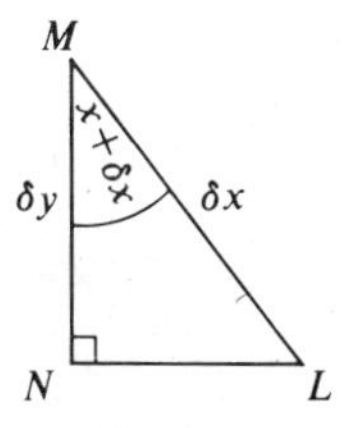

Figure 9.2

Fig. 9.2 is a magnification of the section LMN. When δx is small LM is approximately a straight line, and also the tangent to the circle. The angle between the tangent at M and the vertical is $x + \delta x$. $MN = \delta y$, from the sine curve, and, from triangle LMN:

$$MN = ML.\cos(x+\delta x), \text{ arc } LM = \delta x$$

$$\text{Therefore } MN = \delta x.\cos(x+\delta x)$$

$$\text{Then } \frac{\delta y}{\delta x} = \frac{\delta x.\cos(x+\delta x)}{\delta x} = \cos(x+\delta x)$$

$$\text{As } \delta x \to 0, \frac{\delta y}{\delta x} \to \cos x.$$

$$\text{The derived function} = \frac{dy}{dx} = \cos x \qquad 9.6$$

Figs 9.3 and 9.4 are the corresponding diagrams to Figs 9.1 and 9.2. They refer to the graph of $y = \cos x$ instead of $y = \sin x$.

$$GK = \delta x.\sin(x+\delta x)$$

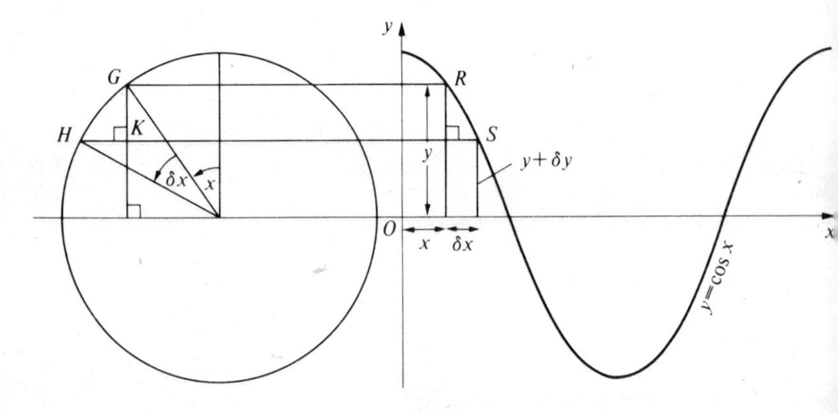

Figure 9.3

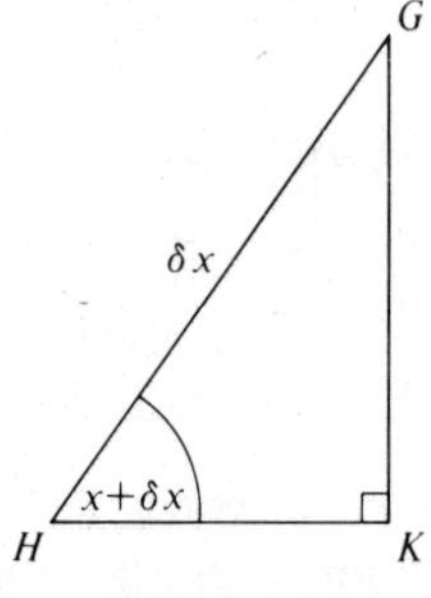

Figure 9.4

$$\frac{\delta y}{\delta x} = \frac{-GK}{\delta x} = \frac{-\delta x \,.\, \sin(x+\delta x)}{\delta x} = -\sin(x+\delta x)$$

$$\text{As } \delta x \to 0, \frac{\delta y}{\delta x} \to -\sin x$$

$$\text{The derived function} = \frac{dy}{dx} = -\sin x \qquad 9.7$$

Graphical methods of estimating the derivatives of sin x and cos x

Fig. 9.5 represents a graph of $y = \sin x$ plotted from $x = 0$ to $x = \pi/2$ radians. $A, B, C, D, E, F, G, H, K$ and L are points on the graph at intervals of 10°. Tangents are drawn as accurately as possible at each of those points. Point P is marked along the negative direction of the x axis so that AP is 1 unit in length. In fact any distance for AP may be chosen. By choosing $AP = 1$ the resulting graph of the gradients is easier to diagnose.

Lines PA_1, PB_1, PC_1 . . . PL_1 are drawn to meet the y axis at A_1, B_1, C_1 . . . L_1. A_1, B_1, C_1 . . . L_1 are projected to the points A_2, B_2, C_2 . . . L_2 above the appropriate displacement along the x axis to give points on the curve which represents the derived function of $\sin x$. From the figure it seems obvious that the derived function is $\cos x$.

By carefully constructing a graph of $y = \cos x$ and applying a similar procedure to the one above it may be shown that the derived function of $y = \cos x$ is $-\sin x$.

Examples

1. The derived function of $a \sin x = \frac{d}{dx}(a \sin x)$

$$= a\frac{d}{dx}(\sin x) \text{ (by 9.4)}$$

$$= a \,.\, \cos x \text{ (by 9.6)}$$

2. The derived function of $b \,.\, \cos x = \frac{d}{dx}(b \,.\, \cos x)$

$$= b \,.\, \frac{d}{dx}(\cos x) \text{ (by 9.4)}$$

$$= b \,.\, -\sin x \text{ (by 9.7)}$$

$$= -b \sin x$$

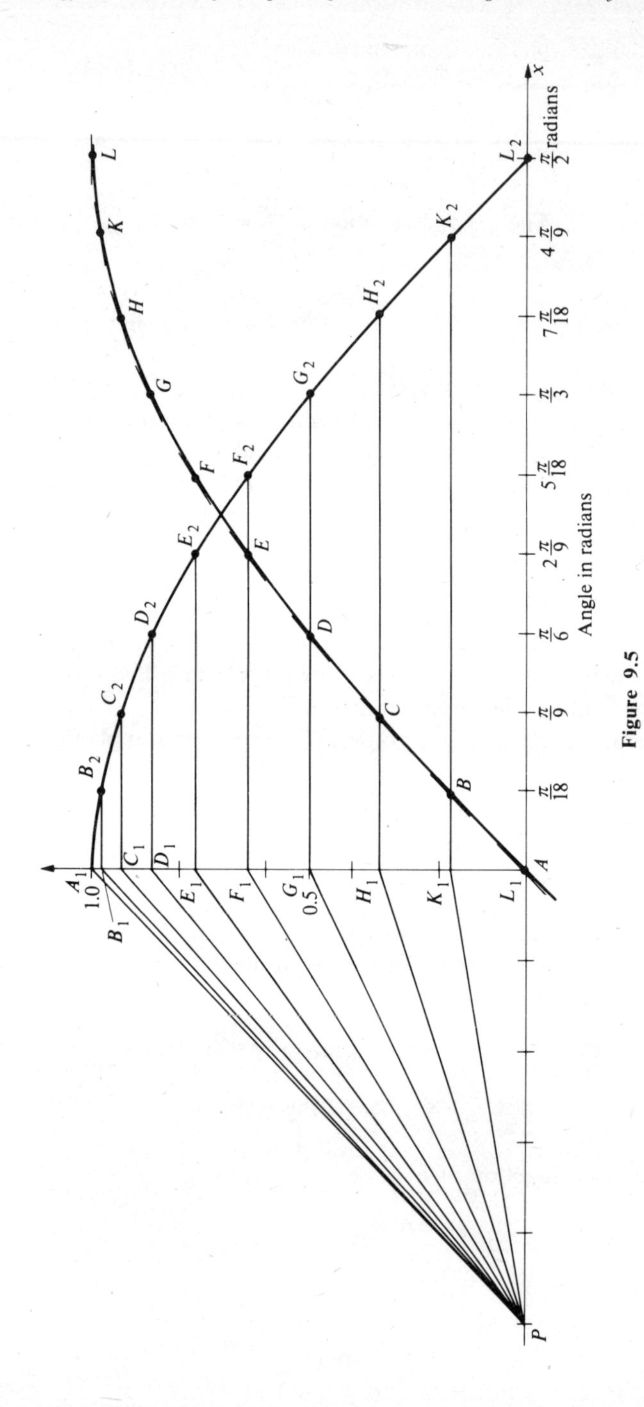

x
radians
Angle in radians
$\frac{\pi}{18}$
$\frac{\pi}{9}$
$\frac{\pi}{6}$
$2\frac{\pi}{9}$
$5\frac{\pi}{18}$
$\frac{\pi}{3}$
$7\frac{\pi}{18}$
$4\frac{\pi}{9}$
$\frac{\pi}{2}$
1.0
0.5
P
A
B
C
D
E
F
G
H
K
L
A_1
B_1
C_1
D_1
E_1
F_1
G_1
H_1
K_1
L_1
B_2
C_2
D_2
E_2
F_2
G_2
H_2
K_2
L_2

Figure 9.5

3. The derived function of $a \cos x + b \sin x$

$$= \frac{d}{dx}(a \cos x + b \sin x)$$

$$= \frac{d}{dx}(a \cos x) + \frac{d}{dx}(b \sin x) \quad \text{(by } 9.2\text{)}$$

$$= a\frac{d}{dx}(\cos x) + b\frac{d}{dx}(\sin x) \quad \text{(by } 9.4\text{)}$$

$$= a . -\sin x + b . \cos x \quad \text{(by } 9.6 \text{ and } 9.7\text{)}$$

$$= b \cos x - a \sin x$$

Exercise 9.2

Determine the derivatives of the following functions w.r.t. the independent variables:

1. $6 \sin x$
2. $-5 \sin x$
3. $\frac{1}{4} \sin x$
4. $-\frac{5}{4} \sin x$
5. $11 \cos x$
6. $-3 \cos x$
7. $\frac{1}{3} \cos x$
8. $-\frac{7}{5} \cos x$
9. $2 \sin x + 3 \cos x$
10. $4 \sin x - 3 \cos x$
11. $-12 \sin x - 13 \cos x$
12. $\frac{1}{4} \cos x - \frac{1}{3} \sin x$
13. $\frac{1}{4}(3 \cos x - 4 \sin x)$
14. $14 - 21 \cos x + 35 \sin x$
15. $\frac{3}{5}(5 \sin x - 6 \cos x)$

Calculate the values of the derivatives of the following functions at the values of the independent variable given:

16. $12 \sin x$: at $x = 0, \pi/4, \pi/2, \pi/3, \pi/6$. (Remember that formulae *9.6* and *9.7* are valid only when the angle is measured in radians.)
17. $\frac{3}{4} \cos x + \frac{5}{8} \sin x$: at $x = 0, \pi/3, \pi/4, \pi/2, 2\pi/3$
18. $5(2 \sin x - 3 \cos x)$: at $x = \pi/6, 3\pi/4, 5\pi/6, 3\pi/2$
19. $8 \sin x + 7 \cos x$: at $x = 45°, 30°, 120°, 180°, 225°$
20. $4 + 2 \cos x - 3 \sin x$: at $x = 90°, 330°, 210°, 240°$

A graphical method for the estimation of derived functions which is an alternative to the one given at the beginning of this section does not depend on the careful construction of tangents to the original curve. That operation is difficult to perform accurately.

The original curve is reproduced accurately by using tracing paper or some other technique. The second curve is displaced a short distance parallel to Ox. The problem with this method is that it is difficult to relate the scale of Oy for the derived curve to that for the original curve. For that reason the method has been omitted.

10

Indefinite and Definite Integrals of Simple Algebraic Functions

10.1 The operation inverse to differentiation

Consider the following pairs of operations:

1. $x+a-a=x$: $(-a)$ is the operation inverse to $+a$ and vice versa.
2. $N\times a\div a=N$: $\div\, a$ is the operation inverse to $\times\, a$ and vice versa.
3. $\sqrt{N^2}=N$: $\sqrt{\ }$ is the operation inverse to the operation of squaring and vice versa.

4. $e^{\ln x}=x$
5. $\ln e^x=x$

(4 and 5:) The operations ln and e to a power are the inverses of each other.

$\dfrac{d}{dx}[f(x)]$ represents the operation of differentiation of $f(x)$. Is there an operation which is the inverse of it? By referring back to Chapter 9:

$$\frac{d}{dx}(ax^n)=a.n.x^{n-1} \qquad \text{(i)}$$

In (i) put $a=1$, $n=1$:

$$\frac{d}{dx}(x)=x^0=1$$

In (i) put $a=\frac{1}{2}$, $n=2$:

$$\frac{d}{dx}(\tfrac{1}{2}x^2)=\tfrac{1}{2}.2x=x$$

In (i) put $a = \frac{1}{3}, n = 3$:

$$\frac{d}{dx}(\tfrac{1}{3}.x^3) = \tfrac{1}{3}.3x^2 = x^2$$

In (i) put $a = \frac{1}{4}, n = 4$:

$$\frac{d}{dx}(\tfrac{1}{4}.x^4) = \tfrac{1}{4}.4x^3 = x^3$$

In (i) put $a = 1/n$:

$$\frac{d}{dx}(\tfrac{1}{n}.x^n) = \tfrac{1}{n}\cdot nx^{n-1} = x^{n-1} \qquad \textit{10.1}$$

Formula *10.1* is a rule for writing down the derivative of the function $\frac{1}{n}.x^n$. The results above can all be written down using that formula and substituting in it special integral values of n.

The operations represented above all involve a movement from left to right: from the function on the left to produce a new function (the derivative) on the right. Suppose we operate this relationship in reverse, i.e. from right to left. Then formula *10.1* tells us that $\frac{1}{n}\cdot x^n$ is the function whose derivative is x^{n-1}. In other words, given x^{n-1}, the derivative, how to determine $\frac{1}{n}x^n$, which might be called the antiderivative. The name given to it is the integral.

To the question 'What is the integral of x^{n-1}?' we answer $\frac{1}{n}x^n$. For the moment we will express this operation:

$$\mathrm{I}(x^{n-1}) = \frac{1}{n}x^n \qquad \textit{10.2}$$

Example

Use formula *10.2* to determine $\mathrm{I}(x^6)$.
Step 1. Put $n-1 = 6$, i.e. $n = 7$.
Step 2. $\mathrm{I}(x^6) = \dfrac{x^n}{n}$, where $n = 7$, i.e. $\dfrac{x^7}{7}$.

This indicates that formula *10.2* could be expressed in a more convenient form. In *10.2* put $n-1 = m$, then $n = m+1$, and:

$$\mathrm{I}(x^m) = \frac{x^{m+1}}{m+1}$$

It is usual to express this using n instead of m:

$$\mathrm{I}\,(x^n) = \frac{x^{n+1}}{n+1} \qquad 10.3$$

In *10.3*, when $n+1=0$, the RHS is not defined because division by zero is not defined. Consequently *10.3* is valid in all cases except when $n=-1$. From *10.3* and *9.5* it appears that:

$$\frac{\mathrm{d}}{\mathrm{d}x}[\mathrm{I}(x^n)] = \frac{\mathrm{d}}{\mathrm{d}x}\left[\frac{x^{n+1}}{n+1}\right] = \frac{1}{n+1}\cdot\frac{\mathrm{d}}{\mathrm{d}x}(x^{n+1}) \text{ (by 9.2)}$$

$$= \frac{1}{n+1}(n+1).x^n \quad \text{(by 9.5)}$$

$$= x^n$$

In other words, I and $\frac{\mathrm{d}}{\mathrm{d}x}$ are inverse operations. To a certain extent they are inverses.

Examples

1. Determine the derivatives w.r.t. x of (a) x^3 and (b) x^3+2.
   (a) $\frac{\mathrm{d}}{\mathrm{d}x}(x^3) = 3x^2$, then $\mathrm{I}\,(3x^2) = x^3$

   (b) $\frac{\mathrm{d}}{\mathrm{d}x}(x^3+2) = 3x^2$, then $\mathrm{I}\,(3x^2) = x^3+2$
2. Determine the derivatives w.r.t. x of (a) x^n and (b) x^n+c.
   (a) $\frac{\mathrm{d}}{\mathrm{d}x}(x^n) = n.x^{n-1}$, then $\mathrm{I}\,(n.x^{n-1}) = x^n$

   (b) $\frac{\mathrm{d}}{\mathrm{d}x}(x^n+c) = n.x^{n-1}$, then $\mathrm{I}\,(n.x^{n-1}) = x^n+c$
3. (a) The derivative of ax^3+bx^2+cx+d, where a, b, c and d are constants, equals $3ax^2+2bx+c$. Then $\mathrm{I}\,(3ax^2+2bx+c) = ax^3+bx^2+cx+d$.

   (b) The derivative of $ax^3+bx^2+cx = 3ax^2+2bx+c$. Therefore $\mathrm{I}(3ax^2+2bx+c) = ax^3+bx^2+cx$.

Examples 1, 2 and 3 above illustrate that the operation of integration does not lead to a unique function. It introduces a constant which may take any value. This operation is called an indefinite integral. The constant is called an arbitrary constant. Consequently formula *10.3* has to be modified to:

$$I(x^n) = \frac{x^{n+1}}{n+1} + c, \text{ when } n+1 \neq 0 \qquad 10.4$$

This is a facet of integration which demonstrates one essential difference between the operations of differentiation and integration. The former provides a unique answer, while the latter does not. Consequently they are not true inverses of each other. For example:

(i) $\frac{d}{dx}[I(x^2)] = \frac{d}{dx}[\tfrac{1}{3}x^3 + c] = x^2$

(ii) $I\left[\frac{d}{dx}(x^2)\right] = I\cdot(2x) = x^2 + c$

i.e. $\frac{d}{dx}[I(x^2)] + c = I\cdot\left[\frac{d}{dx}(x^2)\right]$

Examples

1. $I(x^4) = x^5/5 + c$
2. $I(x^{-3}) = \frac{x^{-2}}{-2} + c = -\frac{1}{2x^2} + c$
3. $I\left(\frac{1}{x^4}\right) = I(x^{-4}) = \frac{x^{-3}}{-3} + c = -\frac{1}{3x^3} + c$
4. $I(\sqrt{x}) = I(x^{\frac{1}{2}}) = \frac{x^{3/2}}{3/2} + c = \frac{2}{3x^{3/2}} + c = \frac{2}{3x\sqrt{x}} + c$

Exercise 10.1

Determine the indefinite integrals of the following functions:

1. x^5
2. x^7
3. x^{10}
4. x^{-2}
5. $1/x^6$
6. x^{-5}
7. $1/x^9$
8. $x^{3/4}$

9. $x^{5/4}$
10. $x^{3/2}$
11. $x^{2\frac{1}{2}}$
12. $x^{2/3}$
13. $x\sqrt{x}$
14. $x \cdot \sqrt[3]{x}$
15. $x^{5/3}$
16. $x^{-\frac{1}{2}}$
17. $x^{-3/2}$
18. $x^{-5/2}$
19. $\dfrac{1}{\sqrt{x}}$
20. $\dfrac{1}{x\sqrt{x}}$
21. $\dfrac{1}{\sqrt[4]{x^5}}$
22. $\dfrac{1}{x^3\sqrt{x}}$
23. $x^2 \times x^3$ (Hint: perform the multiplication, then the integration.)
24. $x^4 \div x^6$
25. $\dfrac{x}{\sqrt{x}}$

Law *10.4* is a law of integration which applies only to functions of a special nature ($x^n, n \neq -1$). It corresponds to the standard forms in differentiation. Does integration conform to general laws similar to *9.2* and *9.4*? The following examples illustrate.

Examples

Using formula *10.4* for appropriate values of n we obtain the following results:

1. When $n = 0$, $\mathrm{I}(1) = \mathrm{I}(x^0) = \dfrac{x^1}{1} + c = x + c$.

2. When $n = 1$, $\mathrm{I}(x) = \mathrm{I}(x^1) = \dfrac{x^2}{2} + c$.

3. When $n = 2$, $\mathrm{I}(x^2) = \dfrac{x^3}{3} + c$.

4. By *9.2* and *9.5*, $\frac{d}{dx}(\frac{1}{3}x^3 + \frac{1}{2}x^2 + x + k) = x^2 + x + 1$.
   Therefore $I(x^2 + x + 1) = \frac{1}{3}x^3 + \frac{1}{2}x^2 + x + k = I(x^2) + I(x) + I(1)$ by 1, 2 and 3 above, because each integral on the right introduces an arbitrary constant itself and these can be adjusted to add together to give k.
5. The result in 4 above may be extended to cover any functions of x which are integrable to give:

$$I[f(x) + g(x) + h(x) + \dots] = I[f(x)] + I[g(x)] + I[h(x)] + \dots \qquad \textit{10.5}$$

6. By *9.4* and *9.5*, $\frac{d}{dx}[\frac{1}{3}ax^3] = \frac{1}{3}a.3x^2 = ax^2$, where a is constant.
   Therefore:

$$\begin{aligned} I(ax^2) &= \tfrac{1}{3}ax^3 + c \\ &= a \cdot \tfrac{1}{3}x^3 + c \\ &= a \cdot I(x^2) \end{aligned}$$

   by 3 above, because $I(x^2)$ introduces a constant of its own and a times that constant can be adjusted to give c.
7. The result in 6 above may be extended to cover any integrable functions of x multiplied by a constant to give:

$$I[a \cdot f(x)] = a . I[f(x)] \qquad \textit{10.6}$$

Formulae *10.5* and *10.6* are the counterparts of *9.2* and *9.4*, for differentiation. They indicate that integration is a linear operator.

Rules for integration of simple algebraic functions

The results above may be summarized in the following three rules:

$$I[f(x) + g(x) + h(x) + \dots] = I[f(x)] + I[g(x)] + I[h(x)] + \dots \quad (\textit{10.5})$$

$$I[a.f(x)] = a.I[f(x)], \text{ where } a \text{ is constant} \quad (\textit{10.6})$$

$$I(x^n) = \frac{x^{n+1}}{n+1} + c,\ n+1 \neq 0 \quad (\textit{10.4})$$

By using the three rules above a fairly wide variety of algebraic functions may be integrated. The following principles must be observed:

1. Only functions which are sums or differences of special terms can be integrated at this stage.

2. The individual terms must be constant multiples of special functions of x.
3. The special functions of x must be expressible in the form x^n, $n+1 \neq 0$.

Examples

Obtain the indefinite integrals of the following functions:

1. $5x^3 - \frac{2}{3}x^2 + 17x - 9$

$\mathrm{I}(5x^3 - \frac{2}{3}x^2 + 17x - 9) = \mathrm{I}(5x^3) + \mathrm{I}(-\frac{2}{3}x^2) + \mathrm{I}(17x) + \mathrm{I}(-9)$ (by *10.5*)

$= 5.\mathrm{I}(x^3) - \frac{2}{3}\mathrm{I}(x^2) + 17\mathrm{I}(x) - 9\mathrm{I}(1)$ (by *10.6*)

$= 5.\frac{1}{4}x^4 - \frac{2}{3}.\frac{1}{3}x^3 + 17.\frac{1}{2}x^2 - 9.x + c$ (by *10.4*)

$= \frac{5}{4}x^4 - \frac{2}{9}x^3 + \frac{17}{2}x^2 - 9x + c$

Check:

$$\frac{\mathrm{d}}{\mathrm{d}x}\left(\frac{5}{4}x^4 - \frac{2}{9}x^3 + \frac{17}{2}x^2 - 9x + c\right)$$

$$= \frac{5}{4}.4x^3 - \frac{2}{9}.3x^2 + \frac{17}{2}.2x - 9 = 5x^3 - \frac{2}{3}x^2 + 17x - 9$$

2. $y = 7x\sqrt{x} - \frac{3}{4}\sqrt{x} + \frac{2}{5}.\frac{1}{\sqrt{x}}$

Step 1. Write $y = 7x^{3/2} - \frac{3}{4}x^{\frac{1}{2}} + \frac{2}{5}x^{-\frac{1}{2}}$.

Step 2.

$$\mathrm{I}(y) = \mathrm{I}(7x^{3/2} - \tfrac{3}{4}x^{\frac{1}{2}} + \tfrac{2}{5}x^{-\frac{1}{2}})$$

$$= \mathrm{I}(7x^{3/2}) + \mathrm{I}(-\tfrac{3}{4}x^{\frac{1}{2}}) + \mathrm{I}(\tfrac{2}{5}x^{-\frac{1}{2}}) \quad \text{(by } \mathit{10.5}\text{)}$$

$$= 7.\mathrm{I}(x^{3/2}) - \tfrac{3}{4}.\mathrm{I}(x^{\frac{1}{2}}) + \tfrac{2}{5}.\mathrm{I}(x^{-\frac{1}{2}}) \quad \text{(by } \mathit{10.6}\text{)}$$

$$= 7.\frac{x^{5/2}}{5/2} - \frac{3}{4}.\frac{x^{3/2}}{3/2} + \frac{2}{5}.\frac{x^{\frac{1}{2}}}{\frac{1}{2}} + c \quad \text{(by } \mathit{10.4}\text{)}$$

$$= \tfrac{14}{5}x^{5/2} - \tfrac{1}{2}x^{3/2} + \tfrac{4}{5}x^{\frac{1}{2}} + c$$

$$= \tfrac{14}{5}x^2\sqrt{x} - \tfrac{1}{2}x\sqrt{x} + \tfrac{4}{5}\sqrt{x} + c$$

Check:

$$\frac{\mathrm{d}}{\mathrm{d}x}\left[\frac{14}{5}x^{5/2} - \tfrac{1}{2}x^{3/2} + \frac{4}{5}x^{\frac{1}{2}} + c\right]$$

$$= \tfrac{14}{5}.\tfrac{5}{2}.x^{3/2} - \tfrac{1}{2}.\tfrac{3}{2}x^{\frac{1}{2}} + \tfrac{4}{5}.\tfrac{1}{2}.x^{-\frac{1}{2}}$$

$$= 7x^{3/2} - \tfrac{3}{4}x^{\frac{1}{2}} + \tfrac{2}{5}x^{-\frac{1}{2}}$$

3. $(x) = (5x+3)(\frac{1}{4}x^2 - x + 1)$

Step 1. Write $f(x) = \frac{5}{4}x^3 - 5x^2 + 5x + \frac{3}{4}x^2 - 3x + 3 = \frac{5}{4}x^3 - \frac{17}{4}x^2 + 2x + 3.$

Step 2.

$$\begin{aligned} &I[f(x)] \\ &= I(\tfrac{5}{4}x^3) + I(-\tfrac{17}{4}x^2) + I(2x) + I(3) \quad \text{(by } \textit{10.5}\text{)} \\ &= \tfrac{5}{4}.\tfrac{1}{4}x^4 - \tfrac{17}{4}.\tfrac{1}{3}x^3 + 2.\tfrac{1}{2}x^2 + 3x + c \quad \text{(by } \textit{10.6} \text{ and } \textit{10.4}\text{)} \\ &= \tfrac{5}{16}x^4 - \tfrac{17}{12}x^3 + x^2 + 3x + c \end{aligned}$$

Check:

$$\begin{aligned} &\frac{d}{dx}\left[\frac{5}{16}x^4 - \frac{17}{12}x^3 + x^2 + 3x + c\right] \\ &= \tfrac{5}{4}x^3 - \tfrac{17}{4}x^2 + 2x + 3 \end{aligned}$$

4. $f(x) = \left[\frac{1}{x^2}(3 + 5\sqrt{x})(1 - 4x^2)\right]$

Step 1. Write:

$$\begin{aligned} f(x) &= \frac{1}{x^2}(3 + 5\sqrt{x} - 12x^2 - 20x^{5/2}) \\ &= \frac{3}{x^2} + \frac{5}{x^{3/2}} - 12 - 20x^{\frac{1}{2}} \\ &= 3.x^{-2} + 5.x^{-3/2} - 12 - 20.x^{\frac{1}{2}} \end{aligned}$$

Step 2.

$$\begin{aligned} &I[f(x)] \\ &= I(3.x^{-2}) + I(5.x^{-3/2}) + I(-12) + I(-20x^{\frac{1}{2}}) \quad \text{(by } \textit{10.5}\text{)} \\ &= 3.I(x^{-2}) + 5.I(x^{-3/2}) - 12.I(1) - 20.I(x^{\frac{1}{2}}) \quad \text{(by } \textit{10.6}\text{)} \\ &= 3.\frac{x^{-1}}{-1} + 5.\frac{x^{-\frac{1}{2}}}{-\frac{1}{2}} - 12.x - 20.\frac{x^{3/2}}{3/2} + c \quad \text{(by } \textit{10.4}\text{)} \\ &= -\frac{3}{x} - \frac{10}{x^{\frac{1}{2}}} - 12x - \frac{40}{3}x^{3/2} + c \\ &= -\frac{3}{x} - \frac{10}{\sqrt{x}} - 12x - \frac{40}{3}x\sqrt{x} + c \end{aligned}$$

Exercise 10.2

Determine the indefinite integrals of the following functions w.r.t. the independent variable.

1. $x^3 - x^2 + x - 1$
2. $x^4 + 2x^3 - 3x^2 + 5x + 2$
3. $x^2 - x + 1 - \dfrac{1}{x^2}$
4. $1 + \dfrac{1}{x^2} - \dfrac{1}{x^3}$
5. $6x^5 - 4x^3$
6. $\dfrac{2}{x^3} - \dfrac{3}{x^4}$
7. $2x^{\frac{1}{2}} + 3x^{-\frac{1}{2}}$
8. $5\sqrt{x} - 2 - \dfrac{2}{\sqrt{x}}$
9. $(x+1)(x+2)$
10. $(2t-3)(3t+4)$
11. $(2+3x-4x^2)(5+7x)$
12. $(2+3x)^2$
13. $(3-2x)^3$
14. $(1+\sqrt{x})^2$
15. $(1+\sqrt{x})^3$
16. $(5+2\sqrt{t})^2$
17. $(6-5t)^3$
18. $\dfrac{1}{t}(t-6t^2)$
19. $\dfrac{1}{x^2}(4+5x^2-3x^4)$
20. $\dfrac{1}{\sqrt{x}}(2+\sqrt{x})(1-\sqrt{x})$
21. $\dfrac{4+2x^2}{3x^2}$
22. $\dfrac{6-3\sqrt{x}+2x}{5\sqrt{x}}$

10.2 One meaning of integration

The derivative of a function can be interpreted as a gradient of a curve, as magnification of a mapping, and as a rate of change of a function. Th following examples lead to one interpretation of integration.

Examples

1. Fig. 10.1 is a sketch of the graph of $y = a$, where a is constant. P is th point (a, x) on $y = a$. The area $OMPL$ (i.e. A) $= a.x = ax$.

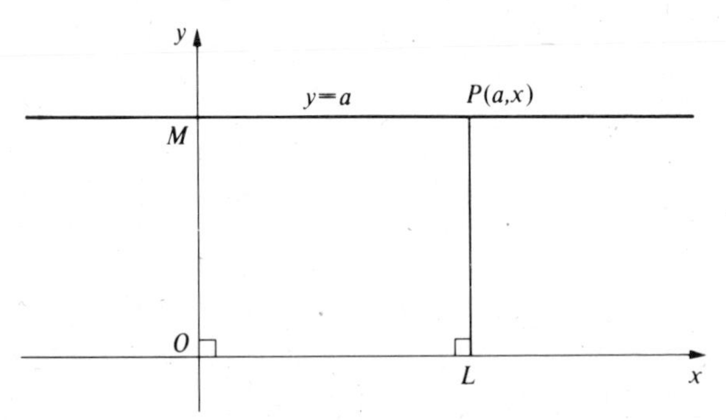

Figure 10.1

Note 1. $\mathrm{I}(a) = a.\mathrm{I}(1) = ax + c$.
Note 2. A and $\mathrm{I}(a)$ differ only by the arbitrary constant.

2. Fig. 10.2 is a sketch of the graph of $y = ax$, where a is constant. P is th point (x, ax) on $y = ax$. The area $OPN\,(A) = \frac{1}{2}ON.NP = \frac{1}{2}x.a$ $= \frac{1}{2}ax^2$.

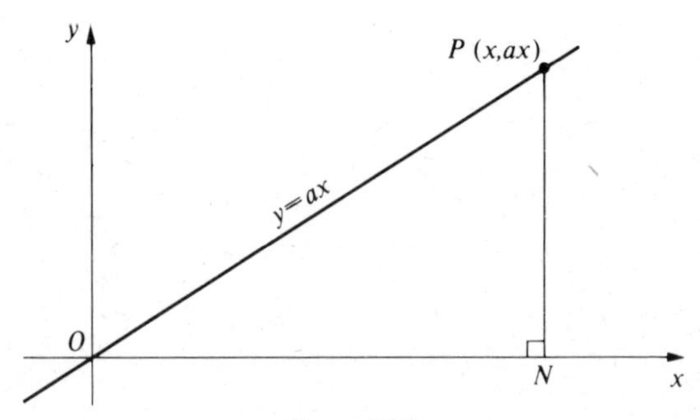

Figure 10.2

Note 1. $I(ax) = a.I(x) = a.\frac{x^2}{2} + c.$

Note 2. A and $I(ax)$ differ only by the arbitrary constant.

3. Fig. 10.3 is a sketch of the graph of $y = ax + b$, where a and b are constant.

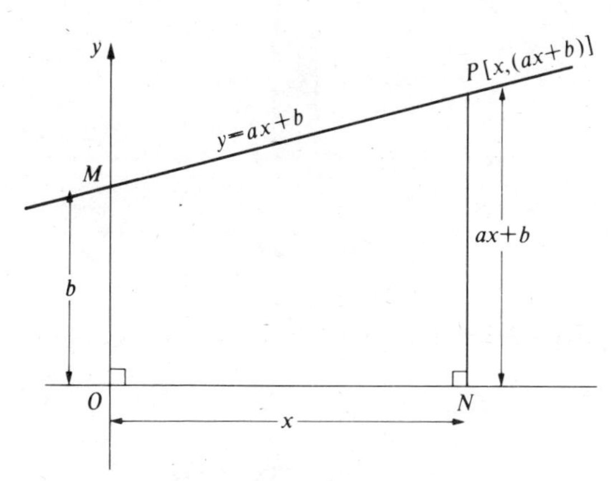

Figure 10.3

P is the point $[x, (ax+b)]$ on $y = ax + b$. The figure $ONPM$ is a trapezium. Area $ONPM$ $(A) = \frac{1}{2}x(b + ax + b) = \frac{1}{2}x(ax + 2b)$ $= \frac{1}{2}ax^2 + bx$.

Note 1.

$$I(ax + b) = I(ax) + I(b)$$
$$= a.I(x) + b.I(1) = a.\tfrac{1}{2}x^2 + b.x + c$$
$$= \tfrac{1}{2}ax^2 + bx + c$$

Note 2. A and $I(ax + b)$ differ only by the arbitrary constant.

If the above examples are typical of more general curves $y = f(x)$ then it appears that $I[f(x)]$ is related to the area under the curve $y = f(x)$.

The area under the curve $y = f(x)$

Fig. 10.4 represents the curve $y = f(x)$. Points P and Q on the curve have co-ordinates (x, y) and $[(x + \delta x), (y + \delta y)]$ respectively. PM and QN are perpendicular to Ox. $PM = y = f(x)$, $QN = f(x + \delta x)$, $MN = \delta x$. Then $PM.MN <$ area $PMNQ < QN.MN$, or $f(x).\delta x <$ area $PQNM$ $< f(x + \delta x).\delta x$.

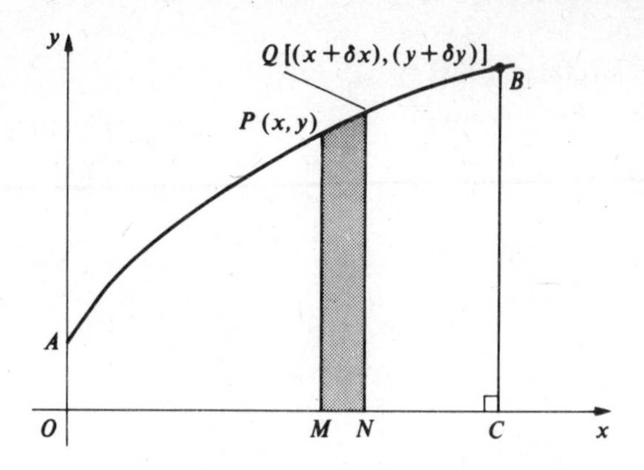

Figure 10.4

Suppose area $OABC$ is A, then area $PQNM$ may be represented by δA. Therefore:

$$f(x).\delta x < \delta A < f(x+\delta x).\delta x$$

$$\therefore \quad f(x) < \frac{\delta A}{\delta x} < f(x+\delta x)$$

As $\delta x \to 0$, $f(x+\delta x) \to f(x)$, providing $f(x)$ is continuous. Therefore $\frac{\delta A}{\delta x}$ $\to f(x)$ as $\delta x \to 0$. That is, $\frac{\delta A}{\delta x} \to$ a limit. That limit is $\frac{dA}{dx}$. Therefore $\frac{dA}{dx}$ $= f(x)$. Consequently $dA = f(x).dx$ and $\delta A \approx f(x).\delta x$. By dividing the area A into a finite number of strips of small width and adding them together we obtain:

$$A = \sum \delta A \approx \sum f(x).\delta x$$

(Note the symbol for summation: $\sum$.) When the increment $\delta x \to 0$ the number of strips increases indefinitely and we change $\sum$ to $\int$, which is an alternative symbol for summation, to arrive at:

$$A = \int f(x).dx \text{ or } \int y.dx \qquad 10.7$$

The following examples illustrate the way this is applied.

Examples

1. From Fig. 10.1 area $OMPL$ is represented by:

$$\int a.dx = ax \qquad \text{(i)}$$

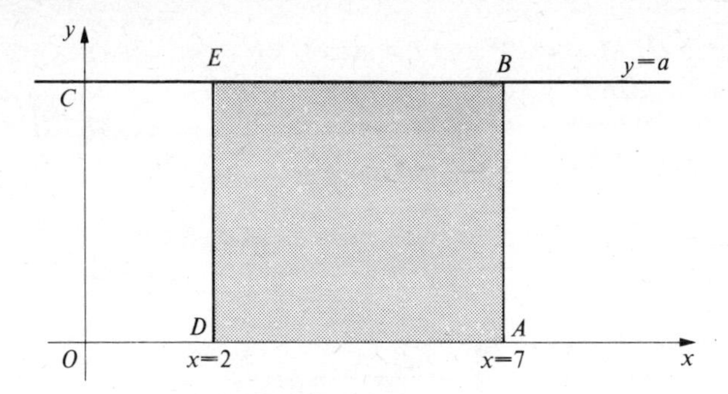

Figure 10.5

(i) is the formula for calculating an area under $y = a$, bounded on the left by $x = 0$.

The shaded area in Fig. 10.5 is bounded not only by the line $y = a$ and Ox but also by the ordinates $x = 2$ and $x = 7$. We cannot apply formula (i) directly to this area because it is not bounded on the left by $x = 0$. Using (i) above, area $OABC = a \times 7$ and area $ODEC = a \times 2$. Area $ABED = OABC - ODEC = a \times 7 - a \times 2 = 7a - 2a = 5a$. As a short cut to the method above we write:

$$ABED = \int_2^7 y.\mathrm{dx} = \int_2^7 a.\mathrm{dx} = [ax]_2^7 = [(a.7) - (a.2)]$$
$$= 7a - 2a = 5a$$

Check: $ABED$ is a rectangle of area $DA \times DE = (7 - 2) \times a = 5a$.

2. Determine the shaded area in Fig. 10.6, i.e. the area $KLMN$.

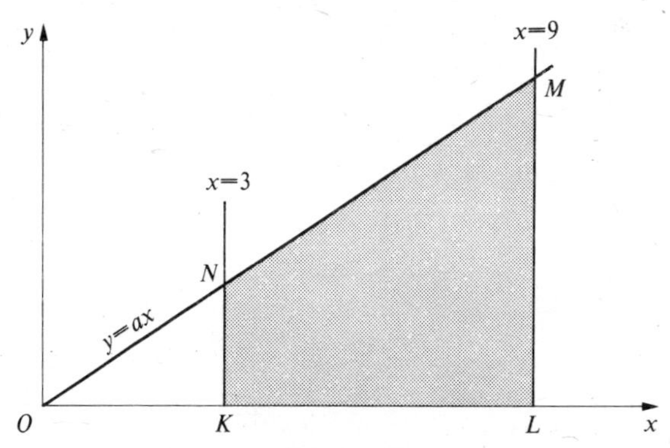

Figure 10.6

From Fig. 10.2 and

$$\int ax.\mathrm{dx} = \tfrac{1}{2}ax^2 + c \qquad \text{(ii)}$$

the area $OLM = \frac{1}{2}a.9^2$ and the area $OKN = \frac{1}{2}a.3^2$. The shaded area $= OLM - OKN = \frac{1}{2}a.9^2 - \frac{1}{2}a.3^2 = \frac{1}{2}a(81-9) = \frac{1}{2}a.72 = 36a$.
Check: $KLMN$ is a trapezium of area $\frac{1}{2}KL(NK+ML)$ $= \frac{1}{2} \times 6(3a+9a) = \frac{1}{2} \times 6 \times 12a = 36a$.

3.

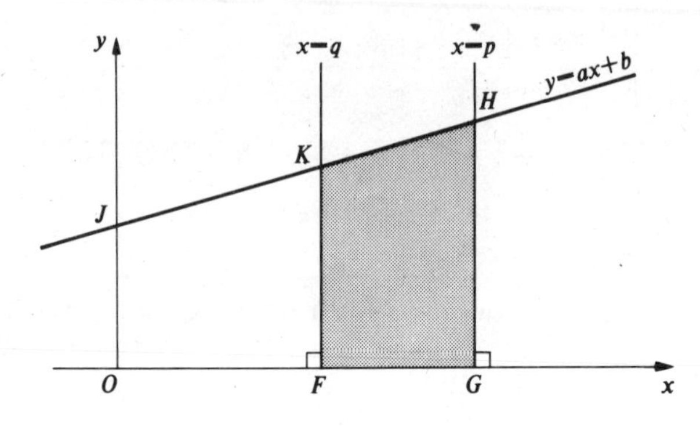

Figure 10.7

Calculate the shaded area in Fig. 10.7. From Fig. 10.3 and the formula there, area $OGHJ = \frac{1}{2}a.p^2 + b.p$, and area $OFKJ = \frac{1}{2}a.q^2 + b.q$. Shaded area $FGHK$ is:

$$(\tfrac{1}{2}ap^2 + bp) - (\tfrac{1}{2}aq^2 + bq)$$
$$= \tfrac{1}{2}a(p^2 - q^2) + b(p-q) = (p-q)\,[\tfrac{1}{2}a(p+q) + b]$$

Alternatively, area $FGHK$ is:

$$\int_q^p y.\mathrm{d}x = \int_q^p (ax+b).\mathrm{d}x$$
$$= [\tfrac{1}{2}ax^2 + bx]_q^p = (\tfrac{1}{2}ap^2 + bp) - (\tfrac{1}{2}aq^2 + bq)$$
$$= (p-q)\,[\tfrac{1}{2}a(p+q) + b]$$

Check: $FGHK$ is a trapezium. Its area is:

$$\tfrac{1}{2} \times FG(FK + GH) = \tfrac{1}{2}(p-q)\,[aq + b + ap + b]$$
$$= \tfrac{1}{2}(p-q)\,[a(p+q) + 2b] = (p-q)\,[\tfrac{1}{2}a(p+q) + b]$$

10.3 Definite integrals

The examples considered above lead to the following definition. The shaded area in Fig. 10.8, under the curve $y = \mathrm{f}(x)$, above Ox and bounded by $x = a$

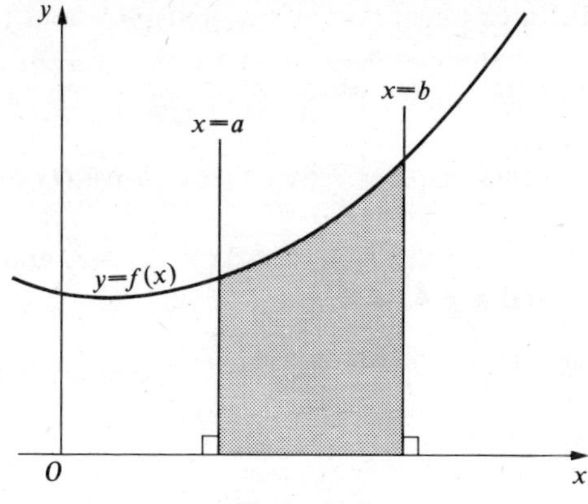

Figure 10.8

and $x = b$, where $a < b$, is equal to:

$$\int_a^b \mathrm{f}(x).\mathrm{d}x$$
$$= [\mathrm{F}(x)]_a^b = \mathrm{F}(b) - \mathrm{F}(a) \qquad 10.8$$

where $\mathrm{F}(x)$ is the indefinite integral of $\mathrm{f}(x)$, i.e. $\mathrm{F}(x) = \int \mathrm{f}(x).\mathrm{d}x$.

Examples

1. Calculate the area under the curve $y = x^2$, above Ox and bounded by the ordinates $x = 1$ and $x = 5$.
   The area $= \int_1^5 y.\mathrm{d}x = \int_1^5 x^2.\mathrm{d}x = [\frac{1}{3}x^3]_1^5 = \frac{1}{3}.125 - \frac{1}{3}.1 = \frac{1}{3}.124 = 41\frac{1}{3}$.
2. Calculate the area under $y = x^3 + 5x^2 + 2x + 1$, above Ox and bounded by the ordinates $x = 2$ and $x = 4$.
   Area $= \int_2^4 y.\mathrm{d}x = \int_2^4 (x^3 + 5x^2 + 2x + 1).\mathrm{d}x$
   $= [\frac{1}{4}x^4 + \frac{5}{3}x^3 + x^2 + x]_2^4 = (\frac{1}{4}.4^4 + \frac{5}{3}.4^3 + 4^2 + 4) - (\frac{1}{4}.2^4 + \frac{5}{3}.2^3$
   $+ 2^2 + 2)$
   $= (64 + \dfrac{320}{3} + 16 + 4) - (4 + \dfrac{40}{3} + 4 + 2)$
   $= 190\frac{2}{3} - 23\frac{1}{3} = 167\frac{1}{3}$

Exercise 10.3

1. Calculate the area under $y = x^2 + 5$, above Ox and bounded by the ordinates $x = 1$ and $x = 3$.

2. Calculate the area under $y = x^2 + x$, above Ox and bounded by the ordinates $x = 2$ and $x = 5$.
3. Calculate the area under $y = x^3 + 4x^2$, above Ox and bounded by $x = 1$ and $x = 6$.
4. Calculate the area under $y = \frac{1}{2}x^3 + \frac{3}{4}x^2$, above Ox and bounded by $x = 2$ and $x = 3$.
5. Calculate the area under $y = x^4 + 3x^3 + 2x + 5$, above Ox and between $x = 1$ and $x = 4$.

Evaluate the following definite integrals:

6. $\int_4^8 x^3 .dx$
7. $\int_3^7 3x^4 .dx$
8. $\int_2^9 (2x^3 + 5x^2 + 3x).dx$
9. $\int_1^6 (\frac{1}{4}x^2 + \frac{1}{3}x + \frac{1}{5}).dx$
10. $\displaystyle\int_2^{10} \left(3 + \frac{1}{x^2}\right).dx$
11. $\displaystyle\int_{1/2}^{3/2} \left(\frac{4x}{5} + \frac{3}{2x^2}\right).dx$
12. $\displaystyle\int_a^{2a} \left(x - \frac{1}{x}\right)^2 .dx$

Negative areas

Sometimes the evaluation of an area by *10.7* or *10.8* produces a negative value. The following examples illustrate this.

Examples

1. Calculate area $\int_2^3 (x^2 - 5x + 4).dx$.
   The shaded area in Fig. 10.9 represents the definite integral. The area

$$= \tfrac{1}{3}x^3 - \tfrac{5}{2}x^2 + 4x$$
$$= (\tfrac{1}{3}.27 - \tfrac{5}{2}.9 + 4.3) - (\tfrac{1}{3}.8 - \tfrac{5}{2}.4 + 4.2)$$
$$= (9 - 22\tfrac{1}{2} + 12) - (2\tfrac{2}{3} - 10 + 8) = -1\tfrac{1}{2} - \tfrac{2}{3} = -2\tfrac{1}{6}$$

   Note: this area is negative because every strip $y.dx$ in $\int y.dx$ involves a $y < 0$ and a $dx > 0$, therefore $y.dx < 0$.

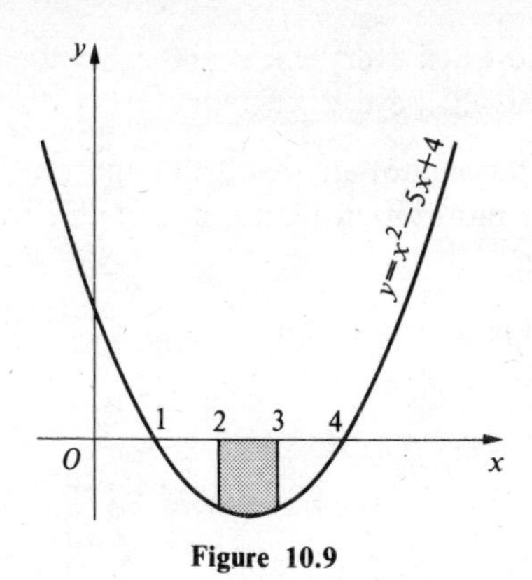

Figure 10.9

2. Calculate area $\int_4^3(-6+7x-x^2).dx$.
   The shaded area in Fig. 10.10 is represented by the definite integral.

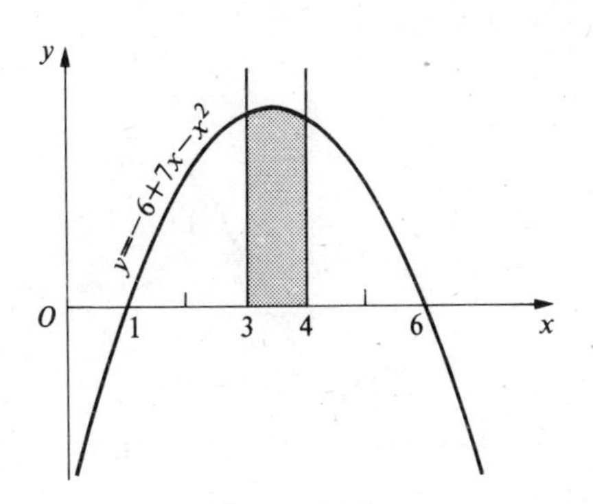

Figure 10.10

The area =

$$\left[-6x+\tfrac{7}{2}x^2-\tfrac{1}{3}x^3\right]_4^3$$
$$=(-6.3+\tfrac{7}{2}.9-\tfrac{1}{3}.27)-(-6.4+\tfrac{7}{2}.16-\tfrac{1}{3}.64)$$
$$=(-18+31\tfrac{1}{2}-9)-(-24+56-21\tfrac{1}{3})$$
$$=4\tfrac{1}{2}-10\tfrac{2}{3}=-6\tfrac{1}{6}$$

Note: this area is negative because for every strip in Fig. 10.10 $y>0$ and $dx<0$, therefore $y.dx<0$.

Inference: The calculation of an area by the definite integral $\int_a^b y.dx$:

(A) will be positive when every $y>0$ and every $dx>0$, i.e. $b>a$
will be positive when every $y<0$ and every $dx<0$, i.e. $b<a$

(B) will be negative when every $y < 0$ and every $dx > 0$, i.e. $b > a$
will be negative when every $y > 0$ and every $dx < 0$, i.e. $b < a$

For most purposes the sign of an area is unimportant. It is the numerical value which matters more often than not. For this reason the following example is important.

3. Calculate area $\int_3^5 (x^2 - 5x + 4).dx$.
The integral =

$$[\tfrac{1}{3}x^3 - \tfrac{5}{2}x^2 + 4x]_3^5$$
$$= (\tfrac{1}{3}.5^3 - \tfrac{5}{2}.5^2 + 4.5) - (\tfrac{1}{3}.3^3 - \tfrac{5}{2}.3^2 + 4.3)$$
$$= (41\tfrac{2}{3} - 62\tfrac{1}{2} + 20) - (9 - 22\tfrac{1}{2} + 12)$$
$$= (-\tfrac{5}{6}) - (-1\tfrac{1}{2}) = -\tfrac{5}{6} + 1\tfrac{1}{2} = \tfrac{2}{3}$$

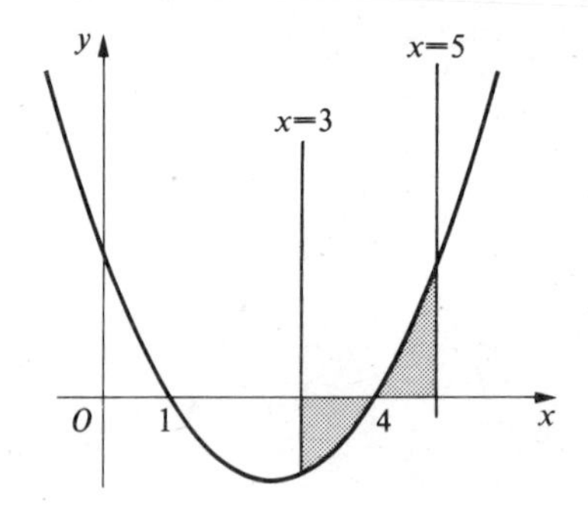

Figure 10.11

The shaded area in Fig. 10.11 can be divided into two parts: from $x = 3$ to $x = 4$, and from $x = 4$ to $x = 5$.
Calculate:

$$\int_3^4 (x^2 - 5x + 4).dx = [\tfrac{1}{3}x^3 - \tfrac{5}{2}x^2 + 4x]_3^4$$
$$- (\tfrac{1}{3}.4^3 - \tfrac{5}{2}.4^2 + 4.4) - (\tfrac{1}{3}.3^3 - \tfrac{5}{2}.3^2 + 4.3)$$
$$= (21\tfrac{1}{3} - 40 + 16) - (9 - 22\tfrac{1}{2} + 12)$$
$$= -2\tfrac{2}{3} + 1\tfrac{1}{2} = -1\tfrac{1}{6}$$

Calculate:

$$\int_4^5 (x^2 - 5x + 4).dx = [\tfrac{1}{3}x^3 - \tfrac{5}{2}x^2 + 4x]_4^5$$
$$= (\tfrac{1}{3}.5^3 - \tfrac{5}{2}.5^2 + 4.5) - (\tfrac{1}{3}.4^3 - \tfrac{5}{2}.4^2 + 4.4)$$
$$= (41\tfrac{2}{3} - 62\tfrac{1}{2} + 20) - (21\tfrac{1}{3} - 40 + 16)$$
$$= (-\tfrac{5}{6}) - (-2\tfrac{2}{3}) = 2\tfrac{2}{3} - \tfrac{5}{6} = 1\tfrac{5}{6}$$

The numerical area $= 1\tfrac{1}{6} + 1\tfrac{5}{6} = 3$

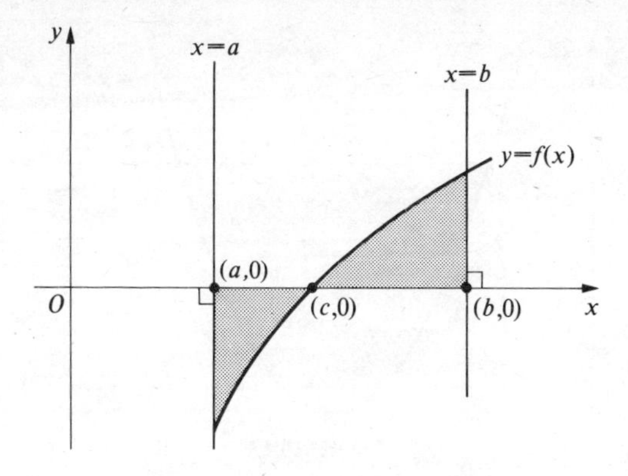

Figure 10.12

Note: Fig. 10.12 represents a case where the curve $y = f(x)$ intersects Ox between $x = a$ and $x = b$, as in Example 3. When the shaded area is required (and sign of area is not taken into account) the area must not be calculated as $\int_a^b f(x).dx$. The two integrals $\int_a^c f(x).dx$ and $\int_c^b f(x).dx$ must be calculated separately. The sum of their numerical values gives the area required.

Whenever the area between $y = f(x)$ and Ox and bounded by $x = a$ and $x = b$ is required it is always advisable to sketch a rough graph to check whether the curve crosses Ox between a and b. If it does then the range a to b must be split up into two or more intervals and the separate integrals evaluated. The examples which follow represent additional problems which may be solved by variations of the standard method.

Examples

1. Calculate the area cut off between $y = x^2 - 5x + 4$ and the x axis. Note: no limits are given for x.
   Fig. 10.13 is a sketch of the curve which crosses Ox at $x = 1$ and $x = 4$. The limits are the ordinates $x = 1$ and $x = 4$.
   Area =

$$\int_1^4 y.dx = \int_1^4 (x^2 - 5x + 4).dx$$
$$= [\tfrac{1}{3}x^3 - \tfrac{5}{2}x^2 + 4x]_1^4$$
$$= (\tfrac{1}{3}.4^3 - \tfrac{5}{2}.4^2 + 4.4) - (\tfrac{1}{3}.1^3 - \tfrac{5}{2}.1^2 + 4.1)$$
$$= (21\tfrac{1}{3} - 40 + 16) - (\tfrac{1}{3} - 2\tfrac{1}{2} + 4) = (-2\tfrac{2}{3}) - (1\tfrac{5}{6}) = -4\tfrac{1}{2}$$

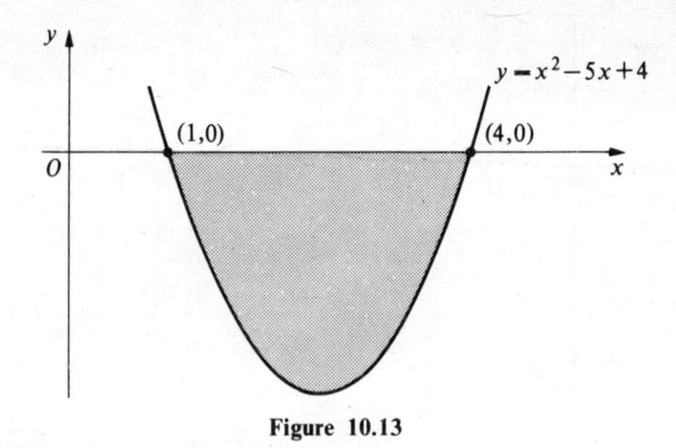

Figure 10.13

2. Calçulate the area enclosed between the curves $x^2 = 4y$ and $y^2 = 4x$. Fig. 10.14 is a sketch of the curves. They intersect at (0, 0) and (4, 4).

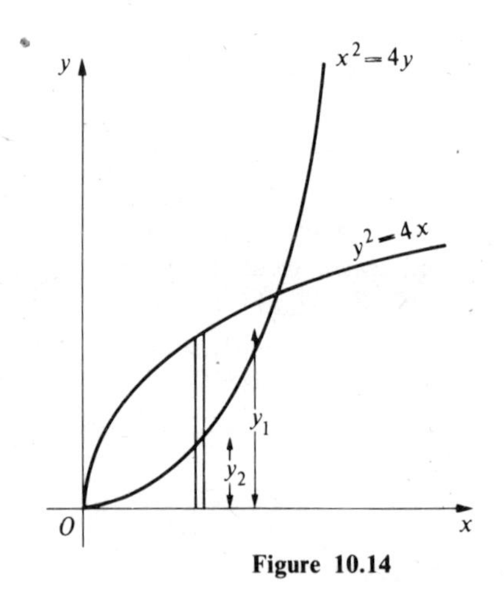

Figure 10.14

These points are obtained by solving the equations of the curves simultaneously. The area enclosed =

$$\int_0^4 (y_1 - y_2).\mathrm{d}x$$

$$= \int_0^4 (2\sqrt{x} - \frac{x^2}{4}).\mathrm{d}x = \left[\frac{4}{3}x^{3/2} - \frac{x^3}{12}\right]_0^4$$

$$= \left(\frac{4 \times 8}{3} - \frac{64}{12}\right) - 0$$

$$= 32/3 - 16/3 = 16/3 = 5\tfrac{1}{3}$$

Exercise 10.4

Calculate the areas cut off between the following curves and the x axis:

1. $y = x^2 - 2x$
2. $y = x^2 - 3x + 2$
3. $y = x^2 + x - 2$
4. $y = x^2 - 7x + 10$
5. $y = x^2 - 3x - 10$
6. $y = 2x^2 - 5x - 3$
7. $y = 3x^2 - 8x + 4$
8. $y = 10 + 3x - x^2$
9. $y = 4 + 3x - x^2$
10. $y = 6 + 10x - 4x^2$

Calculate the areas enclosed between the following pairs of curves:

11. $y^2 = x$ and $y = x$
12. $y^2 = 4x$ and $y = x$
13. $y = x^2 - x$ and $y = x - x^2$
14. $y = x^3$ and $y^2 = x$

ANSWERS TO EXERCISES

Exercise 1.1

1. $(4+a)(x+y)$ **2.** $(a+p)(b+c)$ **3.** $(a+b)(2x+3y)$
4. $(x+2)^2$ **5.** $(x-8)^2$ **6.** $(a+c)(a-c)$ **7.** $2(x+2)^2$
8. $3(x-8)^2$ **9.** $(3-5a)^2$ **10.** $p(a+c)(a-c)$ **11.** $(3a-5)^2$
12. $(7x-2)^2$ **13.** $(9p+4q)(9p-4q)$ **14.** $(x+4)(3x+2)$
15. $(x+3a)(x-a)$ **16.** $(x+a+b)(x+a-b)$

Exercise 1.2

1. $(x+1)(x+2)$ **2.** $(x-1)(x-2)$ **3.** $(x+2)(x-1)$
4. $(x-2)(x+1)$ **5.** $(x+1)(x+4)$ **6.** $(x-1)(x-4)$
7. $(x+4)(x-1)$ **8.** $(x-4)(x+1)$ **9.** $(x-6)(x+1)$
10. $(x+3)(x-2)$ **11.** $(x+2)(x+3)$ **12.** $(x-3)(x+2)$
13. $(x+5)(x-2)$ **14.** $(x-10)(x+3)$ **15.** $(x-8)(x-9)$
16. $(x-12)(x+6)$ **17.** $2(x+4)(x-1)$ **18.** $4(x+4)(x-1)$
19. $11(x+5)(x-2)$ **20.** $a(x+3)(x-2)$

Exercise 1.3

1. $(2x-1)(x-1)$ **2.** $(2x+1)(x+1)$ **3.** $(2x-1)(x+1)$
4. $(2x+1)(x-1)$ **5.** $(3y+1)(y+1)$ **6.** $(3y-1)(y-1)$
7. $(3y-1)(y+1)$ **8.** $(3y+1)(y-1)$ **9.** $(4p+1)(p+1)$
10. $(4p-1)(p-1)$ **11.** $(4p-1)(p+1)$ **12.** $(4p+1)(p-1)$
13. $(2q+1)(q+2)$ **14.** $(2q-1)(q-2)$ **15.** $(2q-1)(q+2)$
16. $(2q+1)(q-2)$ **17.** $(3x+1)(x+2)$ **18.** $(3a-1)(a-2)$
19. $(3b-1)(b+2)$ **20.** $(3c+1)(c-2)$ **21.** $(2x-3)(3x-2)$
22. $(3x-2)(2x+3)$ **23.** $(5y+3)(y+1)$ **24.** $(5x-1)(x-2)$
25. $(7z+6)(z+1)$ **26.** $(7x-3)(x-2)$ **27.** $(7a+2)(a+3)$
28. $(7b-3)(b+2)$ **29.** $(7x+2)(x-3)$ **30.** $(4x+3)(3x+4)$
31. $(4x-3)(3x+4)$ **32.** $2(3x+2)(2x-3)$

Exercise 1.4

The following do not factorize: (2), (4), (5), (6), (7), (8), (9), (10), (11), (12).

Exercise 1.5

1. $(x-1)(x-3)=0$ **2.** $(x-2)(x-3)=0$ **3.** $(x-4)(x-5)=0$
4. $(x-1)(x+1)=0$ **5.** $(x-2)(x+1)=0$ **6.** $(x-3)(x+2)=0$
7. $(x-4)(x+5)=0$ **8.** $(x-a)(x-2a)=0$
9. $(x-a)(x+a)=0$ **10.** $(x-\frac{1}{2})(x-1)=0$

Exercise 1.6

1. (a) $(x-1)(x-\frac{1}{4})=0$ (b) $(x-1)(4x-1)=0$ (c) $x^2-\frac{5}{4}x+\frac{1}{4}=0$
2. (a) $(x-\frac{1}{2})(x-\frac{1}{3})=0$ (b) $(2x-1)(3x-1)=0$ (c) $x^2-\frac{5}{6}x+\frac{1}{6}=0$

3. (a) $(x-\frac{1}{2})(x+1)=0$ (b) $(2x-1)(x+1)=0$ (c) $x^2+\frac{1}{2}x-\frac{1}{2}=0$
4. (a) $(x-\frac{1}{2})(x+\frac{1}{3})=0$ (b) $(2x-1)(3x+1)=0$ (c) $x^2-\frac{1}{6}x-\frac{1}{6}=0$
5. (a) $(x-1\frac{1}{2})(x+3\frac{1}{2})=0$ (b) $(2x-3)(2x+7)=0$ (c) $x^2+2x-\frac{21}{4}=0$
6. (a) $(x+4\frac{1}{2})(x+2\frac{1}{2})=0$ (b) $(2x+9)(2x+5)=0$ (c) $x^2+7x+\frac{45}{4}=0$
7. (a) $(x+3\frac{1}{2})(x+2\frac{1}{4})=0$ (b) $(2x+7)(4x+9)=0$ (c) $x^2+\frac{23}{4}x+\frac{63}{8}=0$
8. (a) $x(x-1)=0$ (b) $x(x-1)=0$ (c) $x^2-x=0$
9. (a) $x(x+2\frac{1}{2})=0$ (b) $x(2x+5)=0$ (c) $x^2+\frac{5}{2}x=0$
10. (a) $(x-a)(x-b)=0$ (b) $(x-a)(x-b)=0$ (c) $x^2-(a+b)x+ab=0$
11. (a) $(x-a)(x+a)=0$ (b) $(x-a)(x+a)=0$ (c) $x^2-a^2=0$
12. (a) $(x+a)^2=0$ (b) $(x+a)^2=0$ (c) $x^2+2ax+a^2=0$
13. (a) $\left(x-\frac{a}{b}\right)\left(x-\frac{2a}{b}\right)=0$ (b) $(bx-a)(bx-2a)=0$

(c) $x^2-\frac{3a}{b}x+\frac{2a^2}{b^2}=0$

14. (a) $(x-a/b)(x-b/a)=0$ (b) $(bx-a)(ax-b)=0$

(c) $x^2-\frac{(a^2+b^2)}{ab}x+1=0$

15. (a) $(x-a)(x-1/a)=0$ (b) $(x-a)(ax-1)=0$

(c) $x^2-\frac{(a^2+1)}{a}x+1=0$

16. (a) $\left(x-\frac{\overline{a+b}}{c}\right)\left(x-\frac{\overline{a-b}}{c}\right)=0$ (b) $(cx-a-b)(cx-a+b)=0$

(c) $x^2-\frac{2a}{c}x+\frac{(a^2-b^2)}{c^2}=0$

Exercise 1.7
The following are quadratic expressions: (1), (2), (3), (4), (5), (9). The following are quadratic equations: (6), (7), (8), (10), (12), (13).

Exercise 1.8
1. $x=2/3$ or $-5/2$ **2.** $x=\frac{1}{4}$ or -4 **3.** $x=0$ or 2 **4.** $x=0$ or 2
5. $x=0$ or $2/5$ **6.** $d=\frac{1}{2}$ or $\frac{4}{3}$ **7.** $s=-1$ or -2 **8.** $p=1$ or $-3/4$
9. $t=2\frac{1}{2}$ or $-1\frac{2}{3}$ **10.** $x=1\frac{2}{3}$ or $-2\frac{1}{2}$

Exercise 1.9
1. 0.851 or −2.351 **2.** 0.207 or −1.207 **3.** 0.679 or −3.679
4. 0.851 or −2.351 **5.** 3.679 or −0.679 **6.** 3.679 or −0.679
7. 0.679 or −3.679 **8.** −0.321 or −4.679 **9.** −0.321 or −4.679
0. 2.151 or 0.349 **11.** 2 or $-\frac{1}{2}$ **12.** 3.646 or −1.646
3. 4.236 or −0.236 **14.** 0.591 or −2.257

Exercise 1.10
. 1.54 s or 6.62 s **2.** 22.93 Ω **3.** 75.3 Ω **4.** 25.50 Ω **5.** 2.04
−11.18 or −5675.1116 **7.** 2.6 m **8.** 0.996 ≈ 1 m **9.** 23.76 A **10.** 0.95

Exercise 1.11

1. $x = 3, y = 5$ or $x = 5, y = 3$ **2.** $x = 6, y = 4$ or $x = -4, y = -6$
3. $x = 3, y = 5$ or $x = -2\frac{1}{2}, y = -6$ **4.** $x = 7, y = -4$ or $x = 4, y = -7$
5. $x = 2.76, y = -4.64$ or $x = -4.19, y = 5.79$
6. $x = 3, y = 4$ or $x = -2, y = -6$ **7.** $x = 5, y = 4$ or $x = -5, y = -4$
8. $x = 1.384, y = -3.768$ or $x = -1.084, y = 1.168$
9. $x = 4, y = 1$ or $x = 3\frac{7}{9}, y = 1\frac{1}{3}$ **10.** $x = 3\frac{1}{2}, y = -\frac{1}{2}$
11. $x = 3\frac{1}{2}, y = \frac{1}{2}$ **12.** $x = 6, y = \frac{1}{2}$ **13.** $x = 5\frac{2}{19}, y = -\frac{3}{19}$
14. $x = 3, y = 2$ **15.** $x = -1, y = -5$ or $x = 3\frac{2}{5}, y = 3\frac{4}{5}$

Exercise 2.1

1. (a) $\log_7 343 = 3$ (b) $\log_{343} 7 = \frac{1}{3}$ (c) $3 \log 7 = \log 343$
(d) $\log_{49} 343 = 3 \log_{49} 7$ (e) $\log_a 343 = 3 \log_a 7$
2. (a) $\log_a c = b$ (b) $\log_c a = 1/b$ (c) $b \log a = \log c$
(d) $\log_{a^2} c = b/2$ (e) $\log_b c = b \log_b a$
3. (a) $\ln N = \ln a + px$ (b) $\log N = \log a + px \log e$
(c) $\log_b N = \log_b a + px \log_b e$ (d) $px \log_{(N/a)} e = 1$
4. (a) $\ln(q/q_0) = -t/CR$ (b) $\log(q/q_0) = -(t/CR)\log e$
(c) $\log_b(q/q_0) = -(t/CR)\log_b e$
5. (a) $\log p + n \log v = \log k$ (b) $\ln p + n \ln v = \ln k$
(c) $\log_v p + n = \log_v k$
6. $125 = 5^3$ **7.** $512 = 2^9$ **8.** $512 = 2^9$
9. $1/64 = 4^{-3}$ **10.** $81 = 3^4$ **11.** $3 = 81^{\frac{1}{4}}$
12. $a = b^c$ **13.** $a^2 = b^p$ **14.** $a = b^{d/c}$ **15.** $a = b^c$
16. $25 = 5^2$ **17.** 3 **18.** 5 **19.** a **20.** a^2
21. a^2 **22.** $1/x$ **23.** N^p **24.** 5 **25.** -3
26. -3 **27.** 3 **28.** 2 **29.** 3 **30.** $-4\frac{1}{6}$
31. 3 **32.** 1 **33.** 5/3 **34.** $\frac{1}{4} \log x$ **35.** $3 + \log_4 25$
36. 6 **37.** $2.523719 \approx 2.524$ **38.** $3.4828921 \approx 3.483$
39. $-4.453131 \approx -4.453$ **40.** $1.2315436 \approx 1.232$
41. $0.8715699 \approx 0.872$ **42.** $7.2636527 \approx 7.264$
43. $4.5628274 \approx 4.563$ **44.** $0.160725 \approx 0.161$
45. $0.1446081 \approx 0.145$ **46.** $1.071991 \approx 1.072$
47. $-0.970181 \approx -0.970$ **48.** 2 or -1 **49.** 2.143 or -0.933
50. $\pm 1.330816 \approx \pm 1.331$ **51.** 2 or -4 **52.** $0.0192372 \approx 0.0192$
53. $4.5775512 \approx 4.578$ **54.** $60.722361 \approx 60.72$
55. $\gamma = 1.1118182 \approx 1.112$; $k = 498.98662 \approx 499.0$

Exercise 3.3

1. (a) 8.18×10^8 s (b) -7.97×10^{-10} kg/s (c) 1.21 kg
2. (a) 9.9 days (b) 1.41 cm/day (c) 26.64 cm
3. (a) 1.9×10^{-4} C (b) -8.0×10^{-3} C/s (c) 1.7×10^{-2} seconds
4. (a) 0.22 A (b) -15.9 amps per second (c) 2.1×10^{-2} seconds
5. (a) 612 N (b) 200 N/rad (c) 0.84 rad
6. (a) 196° C (b) 1.34° C per minute (c) 63 minutes
7. (a) 105.00237 cm (b) 0.0000278 cm per degree C (c) 60 600° C
8. (a) 9.9×10^6 (b) 9.4×10^4 per minute (c) 485 minutes

Exercise 4.1

1. $y = 3.1$; $x = \pm 3.3$ **2.** $y = 37.4$; $x = 2.52$ or 0.48
3. $y = 0.04$; $x = 3.9$ or -0.9 **4.** $y = 1.2$; $x = \pm 1$
5. $y = -0.25$; $x = 3.8$ or 1.2 **6.** Crosses Oy where $y = 1$; $x = -0.27$ or -3.73

7. Axis of symmetry $x = 3/4$, vertex $(3/4, 7/8)$
8. Axis of symmetry $x = 1$, vertex $(1, \frac{1}{2})$
9. vertex $(0, 0)$, $y = x^2$ magnified parallel to Oy by 3
10. vertex $(0, 0)$, $y = x^2$ magnified parallel to Oy by $-4/5$
11. vertex $(1, 0)$, $y = (x-1)^2$ magnified parallel to Oy by 2
12. vertex $(-2, 0)$, $y = -(x+2)^2$ magnified parallel to Oy by $-\frac{1}{3}$
13. vertex $(4, 0)$, $y = (x-4)^2$ magnified parallel to Oy by 2
14. vertex $(0, 3)$, $y = x^2$ translated 3 units along Oy
15. vertex $(0, -6)$, $y = x^2$ translated -6 units along Oy
16. vertex $(0, 5/8)$, $y = x^2$ translated 5/8 units along Oy then magnified by $-3/4$
17. vertex $(2, 5)$, $y = x^2$ translated 5 units along Oy then 2 units parallel to Ox
18. vertex $(-1, -4/3)$, $y = x^2$ translated $-4/3$ units along Oy then -1 unit parallel to Ox, then magnified by $-2/3$
19. vertex $(2/5, 11/6)$, $y = x^2$ translated 11/6 units along Oy then 2/5 units parallel to Ox, then magnified by 4/3
20. vertex (q, r), $y = x^2$ translated r units along Oy then q units along Ox, then magnified by p
21. vertex $(1, 2)$, $y = x^2$ translated 2 units along Oy, then 1 unit along Ox
22. vertex $(-3/2, -12)$, $y = x^2$ translated -12 units along Oy then $-3/2$ units parallel to Ox, then magnified by 4
23. vertex $(\frac{1}{2}, \frac{3}{4})$, $y = x^2$ translated $\frac{3}{4}$ unit along Oy then $\frac{1}{2}$ unit parallel to Ox
24. vertex $(3, 1)$, $x = y^2$ translated 3 units along Ox then 1 unit parallel to Oy, then magnified by 2

Exercise 4.2

1. $y = 1.5$; $x = 0.5$ 2. $y = 0.4$; $x = 0.6$ 3. $y = 5.5$; $x = 2.25$
4. $y = 7.5$; $x = -5.9$ 5. $y = 7.3$; $x = 2.4$
6. $y = -5.2$; $x = 56.6$ 7. $y = 8.05$; $x = 4.5$

Exercise 4.3

1. $y = 18.0$; $x = 6.2$ 2. $y = 58.7$; $x = 8.7$ 3. $y = 60.8$; $x = 8.5$
4. $y = 7.52$; $x = 4.5$ 5. $y = 12.2$; $x = 0.27$ 6. $y = 19$; $x = 0.244$
7. $y = 48.8$; $x = 0.156$ 8. $y = 58.7$; $x = 1.52$

Exercise 5.1

1. $y = 138\,000$; $x = 10.63$ 2. $y = 20.41$; $x = 0.224$
3. $y = 87.39$; $x = 4.27$ 4. $k = 0.000125$; $W = 14.45$ kg; $R = 518$ mm
5. $a = 1.4$; $V = 8.52$ m/s; $h = 77.2$ m
6. $I = 0.004$ V; $I = 0.242$ A; $V = 162.5$ V
7. $T = 2.L^{\frac{1}{2}}$; $L = 6.25$ m; $T = 3.162$ s
8. $t = 42.V^{-1}$; $t = 1.87$ s; $V = 14$ m/s
9. $y = 8.95x^{0.423}$; $y = 11.6$; $x = 0.62$
10. $N \approx 8300 \times d^{-1.07}$ N: 260.6, 168.8; d: 25.45, 36.85
11. $a \approx 200$; $n \approx -2$ F: 8, 4.08; d: 2.5, 6.2
12. $P \approx 42000 \times v^{-1.015}$ P: 3320, 4730; v: 13.0, 8.3
13. $a = 0.52$; $n = -3$ W: 0.005; $V = 7.5$

Exercise 6.1

1. $\begin{pmatrix} 3 & 6 \\ 6 & 5 \end{pmatrix}$ 2. $\begin{pmatrix} 15 & 24 \\ 21 & 6 \end{pmatrix}$ 3. $\begin{pmatrix} 9 & 7 \\ 5 & 8 \end{pmatrix}$ 4. $\begin{pmatrix} 8 & 14 \\ 5 & -2 \end{pmatrix}$

5. $\begin{pmatrix} 3a & 11a \\ 9a & 17a \end{pmatrix}$ 6. $\begin{pmatrix} 6b & 5b \\ -2b & -16b \end{pmatrix}$ 7. $\begin{pmatrix} \overline{2a-b} & \overline{3a+2b} \\ \overline{5a-b} & \overline{a+5b} \end{pmatrix}$

8. $\begin{pmatrix} 1 & 1 \\ 1 & 1 \end{pmatrix}$ 9. $\begin{pmatrix} 6 & 1 \\ 3 & 6 \end{pmatrix}$ 10. $\begin{pmatrix} 19 & -5 \\ -13 & -13 \end{pmatrix}$

11. $\begin{pmatrix} 62 & 7 \\ 31 & 80 \end{pmatrix}$ **12.** $\begin{pmatrix} 73 & -54 \\ 12 & -72 \end{pmatrix}$

13. $\begin{pmatrix} 11a & 11a \\ -22a & 22a \end{pmatrix}$ **14.** $\begin{pmatrix} -a+9b & -2a+5b \\ 4a-b & -3b-5a \end{pmatrix}$

Exercise 6.2

1. $\begin{pmatrix} 7 & 8 \\ 18 & 17 \end{pmatrix}$ **2.** $\begin{pmatrix} -2 & 38 \\ -6 & 54 \end{pmatrix}$ **3.** $\begin{pmatrix} 13 & -4 \\ 4 & -1 \end{pmatrix}$

4. $\begin{pmatrix} 5a & -b \\ -c & 5d \end{pmatrix}$ **5.** $\begin{pmatrix} pa+r & -qa+r \\ qa+r & pa \end{pmatrix}$ **6.** $\begin{pmatrix} 3ax & 7bx \\ -7ay & 3by \end{pmatrix}$

7. $\begin{pmatrix} 1 & 0 \\ 0 & 1 \end{pmatrix}$ **8.** $\begin{pmatrix} 1 & 0 \\ 0 & 1 \end{pmatrix}$ **9.** $\begin{pmatrix} 1 & 0 \\ 0 & 1 \end{pmatrix}$ **10.** $\begin{pmatrix} 1 & 0 \\ 0 & 1 \end{pmatrix}$

Exercise 6.3

1. $\begin{pmatrix} 4 \\ 10 \end{pmatrix}$ **2.** $\begin{pmatrix} 23 \\ 3 \end{pmatrix}$ **3.** $\begin{pmatrix} -14 \\ 1 \end{pmatrix}$ **4.** $\begin{pmatrix} 9 \\ -29 \end{pmatrix}$ **5.** $\begin{pmatrix} a \\ b \end{pmatrix}$

6. $\begin{pmatrix} -a \\ -3a \end{pmatrix}$ **7.** $\begin{pmatrix} -3a^2+4b^2 \\ -5a^2-7b^2 \end{pmatrix}$ **8.** $\begin{pmatrix} aA+bB \\ cA+dB \end{pmatrix}$ **9.** $\begin{pmatrix} aC+bD \\ cC+dD \end{pmatrix}$

Exercise 6.4

1. $\begin{pmatrix} 5 & 11 \\ 11 & 25 \end{pmatrix}$ **2.** $\begin{pmatrix} 5 & 4 \\ 4 & 5 \end{pmatrix}$ **3.** $\begin{pmatrix} 5 & 4 \\ 4 & 5 \end{pmatrix}$ **4.** $\begin{pmatrix} 50 & 82 \\ -7 & -5 \end{pmatrix}$

5. $\begin{pmatrix} 36 & 0 \\ 39 & 9 \end{pmatrix}$ **6.** $\begin{pmatrix} -28 & 13 \\ 21 & -6 \end{pmatrix}$ **7.** $\begin{pmatrix} -34 & 7 \\ 15 & 0 \end{pmatrix}$ **8.** $\begin{pmatrix} ac+bd & ad+bc \\ bc+ad & bd+ac \end{pmatrix}$

9. $(A \times B) \times C = A \times (B \times C) = \begin{pmatrix} 59 & 27 \\ 133 & 61 \end{pmatrix}$

10. $\begin{pmatrix} -93 & 141 \\ 224 & -192 \end{pmatrix}$ **11.** $\begin{pmatrix} -351 & -26 \\ 363 & 66 \end{pmatrix}$

12. $\begin{pmatrix} ac+bd & ad+bc \\ bc+ad & bd+ac \end{pmatrix}$ **13.** $\begin{pmatrix} ca+db & cb+da \\ da+cb & db+ca \end{pmatrix}$

14. $\begin{pmatrix} aA+bB & aC+bD \\ cA+dB & cC+dD \end{pmatrix}$ **15.** $\begin{pmatrix} Aa+Cc & Ab+Cd \\ Ba+Dc & Bb+Dd \end{pmatrix}$

16. $\begin{pmatrix} 3 & 5 \\ -2 & 7 \end{pmatrix}$ **17.** $\begin{pmatrix} -18 & -35 \\ 29 & -17 \end{pmatrix}$ **18.** $\begin{pmatrix} 3 & 5 \\ -2 & 7 \end{pmatrix}$

19. $\begin{pmatrix} -18 & -35 \\ 29 & -17 \end{pmatrix}$ **20.** $\begin{pmatrix} a & b \\ c & d \end{pmatrix}$ **21.** $\begin{pmatrix} a & b \\ c & d \end{pmatrix}$

Exercise 6.5

1. $\begin{pmatrix} 1 & 0 \\ 0 & 1 \end{pmatrix}$ **2.** $\begin{pmatrix} 1 & 0 \\ 0 & 1 \end{pmatrix}$ **3.** $\begin{pmatrix} 1 & 0 \\ 0 & 1 \end{pmatrix}$ **4.** $\begin{pmatrix} 1 & 0 \\ 0 & 1 \end{pmatrix}$

5. $\begin{pmatrix} 1 & 0 \\ 0 & 1 \end{pmatrix}$ **6.** $\begin{pmatrix} 1 & 0 \\ 0 & 1 \end{pmatrix}$ **7.** $\begin{pmatrix} 5 & 0 \\ 0 & 5 \end{pmatrix}$ **8.** $\begin{pmatrix} 5 & 0 \\ 0 & 5 \end{pmatrix}$

9. $\begin{pmatrix} 8 & 0 \\ 0 & 8 \end{pmatrix}$ **10.** $\begin{pmatrix} 8 & 0 \\ 0 & 8 \end{pmatrix}$ **11.** $\begin{pmatrix} 11 & 0 \\ 0 & 11 \end{pmatrix}$ **12.** $\begin{pmatrix} 11 & 0 \\ 0 & 11 \end{pmatrix}$

Exercise 6.6

1. 1 **2.** 1 **3.** 1 **4.** 5 **5.** 5 **6.** 1 **7.** 8 **8.** 8
9. -44 **10.** 32 **11.** $2a^2$ **12.** a^2 **13.** 2 **14.** a^2 **15.** $3a$

Exercise 7.3

1. $\begin{pmatrix} 2 & -1 \\ -5/2 & 3/2 \end{pmatrix}$ **2.** $\begin{pmatrix} 2/5 & -1/5 \\ -3/5 & 4/5 \end{pmatrix}$ **3.** $\begin{pmatrix} 2/5 & -3/5 \\ -3/5 & 7/5 \end{pmatrix}$

4. $\begin{pmatrix} 4/21 & -1/7 \\ -5/21 & 3/7 \end{pmatrix}$ **5.** $\begin{pmatrix} 3/22 & -5/22 \\ 1/11 & 2/11 \end{pmatrix}$ **6.** $\begin{pmatrix} 1 & -3/2 \\ 2 & -5/2 \end{pmatrix}$ **7.** $\begin{pmatrix} 3/5 & -2/5 \\ 4/5 & -1/5 \end{pmatrix}$

8. $\begin{pmatrix} -3/23 & -2/23 \\ 7/23 & -3/23 \end{pmatrix}$ **9.** $\begin{pmatrix} 1/2a & -1/2a \\ 1/2b & 1/2b \end{pmatrix}$ **10.** $\begin{pmatrix} -5/97 & 12/97 \\ 11/97 & -7/97 \end{pmatrix}$

Exercise 7.4

1. $x = 7, y = -2$ **2.** $x = 3, y = 1$ **3.** $x = -3\frac{4}{15}, y = \frac{3}{5}$ **4.** $a = 3, b = 1$
5. $x = 13, y = -20$ **6.** $p = -23/39, q = 5/13$ **7.** $x = 2\frac{3}{5}, y = 1\frac{19}{20}$

Exercise 7.5

8. $x = 8, y = 3$ **9** $x = 3/2, y = 3/4$ **10.** $x = 2/7, y = 1$
No solutions for 11, 12, 13, 14, 15, 16

Exercise 7.6

The following are singular matrices: 1, 3, 4, 5, 7.
9. $p = \pm 1$ **10.** $q = 3\frac{3}{4}$ **11.** $p = -8\frac{3}{4}$
12. $p = \pm 9$ **13.** $a = 5$ **14.** $a = -5$ **15.** $x + 3y = 0$
16. $p + 6 \neq 0$ **17.** $5a + 6 \neq 0$ **18.** $a \neq 0$ and $b \neq 0$

Exercise 7.7

1. $i_1 = 1\frac{32}{35}$; $i_2 = 1\frac{31}{70}$ **2.** $V_1 = 46$; $V_2 = 34$
3. $i_1 = -9/46$; $i_2 = 2\frac{47}{184}$ **4.** $a = 1/15$; $b = 83\frac{1}{3}$
5. $p = 2894\frac{14}{19}$; $q = 5/19$ **6.** $V_1 = 108\frac{1}{29}$; $V_2 = 35\frac{1}{29}$
7. $r_1 = 5\frac{325}{384}$; $r_2 = 5\frac{17}{192}$

Exercise 8.1

1. 3/2; 56.31° **2.** $-1/2$; 153.43° **3.** 2/5; 21.8°
4. $-3/4$; 143.13° **5.** -2; 116.57° **6.** 5/3; 59.04°
7. 1/9; 6.34° **8.** $-6/13$; 156.22° **9.** 2; 63.43°

10. 2.1 **11.** 2.01 **12.** k/h
13. $2+h$ **14.** 4.1 **15.** 4.01

Exercise 8.2
1. grad = mag = 2 **2.** grad = mag = $-\frac{1}{4}$ **3.** -1 **4.** 3 **5.** 2/3
6. 2 **7.** $-4/5$ **8.** -2 **9.** $-a$ **10.** $-b/a$ **11.** $-a/b$
12. $-3/2$ **13.** $-b/a$ **14.** $-ad/bc$

Exercise 8.3
1. 2.5 **2.** 2.1 **3.** 2.01 **4.** 4.5 **5.** 4.1
6. 4.01 **7.** 3 **8.** 2.1 **9.** 3.1 **10.** 3.01 **11.** $4+h$
12. $6+h$ **13.** $4+h$ **14.** $6+h$ **15.** $5+h$ **16.** $7+h$

Exercise 8.4
1. 8 **2.** 12 **3.** -4 **4.** 10 **5.** 20 **6.** -6 **7.** -16
8. 8 **9.** 2 **10.** 3 **11.** 4 **12.** 0 **13.** 0 **14.** 3

Exercise 8.5
1. $12x$ **2.** $-8x$ **3.** $4x/3$ **4.** $2x+1$ **5.** $2x+3$
6. $2x-1$ **7.** $2ax-b$ **8.** $2ax/b$ **9.** $2px-q$

Exercise 8.6
1. $12x$; $0, 12, -24, 36, -48, 12a$ **2.** $-8x$; $0, -4, 2, -8/3, 16, -8b$
3. $4x/3$; $-8/3, -4, 16/3, 20/3, -1/3, 4a/3$
4. $2x+1$; $1, -3, 9, -11, 17, 2a+1$ **5.** $2x+3$; $3, 5, -1, 13/3, 1/2$

Exercise 9.1
1. $8x^7$ **2.** $10x^9$ **3.** $12x^{11}$ **4.** $-3x^{-4}$
5. $-8x^{-9}$ **6.** $-2x^{-3} = -2/x^3$ **7.** $-4x^{-5} = -4/x^5$
8. $-6x^{-7} = -6/x^7$ **9.** $\frac{1}{2}x^{-1/2}$ **10.** $\frac{3}{2}x^{1/2}$ **11.** $\frac{5}{2}x^{3/2}$

12. $-\frac{1}{2}x^{-3/2}$ 13. $-\frac{3}{2}x^{-5/2}$ **14.** $-\frac{5}{4}x^{-9/4}$ **15.** $\dfrac{1}{2\sqrt{x}}$

16. $3\sqrt{x}/2$ **17.** $-5/2u^3\sqrt{u}$ **18.** $-\frac{3}{4}u^{-7/4}$ **19.** $\frac{2}{3}y^{-1/3}$
20. $\frac{3}{2}\sqrt{y}$ **21.** $22x$ **22.** $-\frac{8}{3}x^3$

23. $-\dfrac{15}{16x^2}$ **24.** $-\dfrac{12}{5x^3}$ **25.** $\dfrac{3}{8\sqrt{x}}$ **26.** $\frac{1}{2}x^{1/7}$

27. $\dfrac{49}{6x^4\sqrt{x}}$ **28.** $2x+5$ **29.** $6x-7$ **30.** $4-3/x^2$
31. $11+4/x^3$ **32.** $6x+5+18/x^3$ **33.** $2x-2$: 4
34. $8x+20$: 12 **35.** $8x-25$: -17 **36.** $9x^2-10x+6$: 62
37. $24x^2-24x+6$: 54 **38.** $-1/x^2$: -1 **39.** $-5/3x^2$: $-5/12$

40. $\dfrac{6}{5}-\dfrac{7}{5x^2}$: 47/45 **41.** $1/5-4/5x^2$: -3

42. $\dfrac{3}{2}+\dfrac{1}{2u^2}+\dfrac{5}{2u^3}$: $-\dfrac{1}{2}$ **43.** $-\dfrac{1}{2x^{3/2}}-\dfrac{1}{x^2}$: $-\dfrac{1}{8}$

44. $-1/2x^{3/2} - 2/x^2 + 9/2x^{5/2} + 12/x^3$: 6
45. $24x^2 - 10x + 11$: 197 **46.** $32x^3 + 6x^2 - 56x - 11$: 157
47. $63x^2 - 30x - 27 + 39/x^2$: $174\frac{3}{4}$
48. $24 + 30/x^2 - 42/x^4 + 50/x^6$: 62
49. $-15/x^4 - 2/x^2 - 15 - 18x^2$: $-88\frac{7}{16}$
50. $480x^3 - 162x^2 - 330x - 36$: 2496

Exercise 9.2

1. $6\cos x$ **2.** $-5\cos x$ **3.** $\frac{1}{4}\cos x$
4. $-\frac{5}{4}\cos x$ **5.** $-11\sin x$ **6.** $3\sin x$
7. $-\frac{1}{3}\sin x$ **8.** $\frac{7}{5}\sin x$ **9.** $2\cos x - 3\sin x$
10. $4\cos x + 3\sin x$ **11.** $-12\cos x + 13\sin x$ **12.** $-\frac{1}{4}\sin x - \frac{1}{3}\cos x$
13. $-\frac{1}{4}(3\sin x + 4\cos x)$ **14.** $21\sin x + 35\cos x$
15. $\frac{3}{5}(5\cos x + 6\sin x)$ **16.** $12, 6\sqrt{2}, 0, 6, 6\sqrt{3}$
17. $5/8, (5/16 - 3\sqrt{3}/4), 11\sqrt{2}/16, -3/4, -3\sqrt{3}/8, -5/16$
18. $\frac{5}{2}(2\sqrt{3}+3), 5\sqrt{2}/2, \frac{5}{2}(3-2\sqrt{3}), -15$
19. $\sqrt{2}/2, \frac{1}{2}(8\sqrt{3}-7), (4-(7\sqrt{3}/2)), -8, -\sqrt{2}/2$
20. $-2, \left(1 - \frac{3\sqrt{3}}{2}\right), \left(1 + \frac{3\sqrt{3}}{2}\right), (\sqrt{3} + 3/2)$

Exercise 10.1

The arbitrary constant has been omitted in each of the following answers.

1. $\frac{1}{6}x^6$ **2.** $\frac{1}{8}x^8$ **3.** $\frac{1}{11}x^{11}$ **4.** $-x^{-1}$ **5.** $-\frac{1}{5}x^{-5}$
6. $-\frac{1}{4}x^{-4}$ **7.** $-\frac{1}{8x^8}$ **8.** $\frac{4}{7}x^{7/4}$ **9.** $\frac{4}{9}x^{9/4}$ **10.** $\frac{2}{5}x^{5/2}$
11. $\frac{2}{7}x^{7/2}$ **12.** $\frac{3}{5}x^{5/3}$ **13.** $\frac{2}{5}x^2\sqrt{x}$ **14.** $\frac{3}{7}x^2\sqrt{x}$ **15.** $\frac{3}{8}x^{8/3}$
16. $2x^{1/2}$ **17.** $-2x^{-1/2}$ **18.** $-\frac{2}{3}x^{-3/2}$ **19.** $2\sqrt{x}$ **20.** $-2/\sqrt{x}$
21. $-4/\sqrt[4]{x}$
22. $-2/5x^2\sqrt{x}$ **23.** $\frac{1}{6}x^6$ **24.** $-1/x$ **25.** $\frac{2}{3}x\sqrt{x}$

Exercise 10.2

The arbitrary constant has been omitted in each of the following answers.

1. $\frac{1}{4}x^4 - \frac{1}{3}x^3 + \frac{1}{2}x^2 - x$ **2.** $\frac{1}{5}x^5 + \frac{1}{2}x^4 - x^3 + \frac{5}{2}x^2 + 2x$
3. $\frac{1}{3}x^3 - \frac{1}{2}x^2 + x + 1/x$ **4.** $x - 1/x + 1/2x^2$
5. $x^6 - x^4$ **6.** $-1/x^2 + 1/x^3$
7. $\frac{4}{3}x^{3/2} + 6x^{1/2}$ **8.** $\frac{10}{3}x\sqrt{x} - 2x - 4\sqrt{x}$
9. $\frac{1}{3}x^3 + \frac{3}{2}x^2 + 2x$ **10.** $2t^3 - \frac{1}{2}t^2 - 12t$
11. $10x + \frac{29}{2}x^2 + \frac{1}{3}x^3 - 7x^4$ **12.** $4x + 6x^2 + 3x^3$
13. $27x - 27x^2 + 12x^3 - 2x^4$ **14.** $x + \frac{4}{3}x\sqrt{x} + \frac{1}{2}x^2$
15. $x + 2x\sqrt{x} + \frac{3}{2}x^2 + \frac{2}{5}x^2\sqrt{x}$ **16.** $25t + \frac{40}{3}t\sqrt{t} + 2t^2$
17. $216t - 270t^2 + 150t^3 - \frac{125}{4}t^4$ **18.** $t - 3t^2$
19. $-4/x + 5x - x^3$ **20.** $4\sqrt{x} - x - \frac{2}{3}x\sqrt{x}$
21. $-4/3x + 2x/3$ **22.** $\frac{12}{5}\sqrt{x} - \frac{3}{5}x + \frac{4}{15}x\sqrt{x}$

Exercise 10.3

1. $18\frac{2}{3}$ **2.** $49\frac{1}{2}$ **3.** $610\frac{5}{12}$ **4.** $12\frac{7}{8}$ **5.** $425\frac{17}{20}$ **6.** 960
7. $9938\frac{2}{5}$ **8.** $4589\frac{2}{3}$ **9.** $24\frac{3}{4}$ **10.** $24\frac{2}{5}$ **11.** $2\frac{4}{5}$ **12.** $\frac{7}{3}a^3 - \frac{2}{3}a + \frac{1}{2a}$

Exercise 10.4

1. $-1\frac{1}{3}$ **2.** $-1/6$ **3.** $-4\frac{1}{2}$ **4.** $-4\frac{1}{2}$ **5.** $-57\frac{1}{6}$
6. $-14\frac{7}{24}$ **7.** $-1\frac{5}{27}$ **8.** $57\frac{1}{6}$ **9.** $20\frac{5}{6}$
10. $28\frac{7}{12}$ **11.** 1/6 **12.** $2\frac{2}{3}$ **13.** 1/3 **14.** 5/12

Index